THE UNIVERSE AND BEYOND

FOURTH EDITION

TERENCE DICKINSON

Foreword by Skylab Astronaut Edward G. Gibson

A FIREFLY BOOK

Published by Firefly Books Ltd. 2004

Fourth Edition, 2004

Second Printing, 2006

Publisher Cataloging-in-Publication Data (U.S.)

Dickinson, Terence
The universe and beyond / Terence Dickinson ;
foreword by Edward G. Gibson.
4th ed.
[180] p. : col. ill., photos., charts, maps ; cm.
Includes bibliographical references and index.

ISBN-13: 978-1-55297-937-2 (bound)
ISBN-10: 1-55297-937-7 (bound)
ISBN-13: 978-1-55297-901-3 (pbk.)
ISBN-10: 1-55297-901-6 (pbk.)

1. Astronomy. I. Title.
520 21 QB44.2.D495 2004

Published in the United States by
Firefly Books (U.S.) Inc.
P.O. Box 1338, Ellicott Station
Buffalo, New York, 14205

Produced by
Bookmakers Press Inc.
12 Pine Street
Kingston, Ontario K7K 1W1
(613) 549-4347
tcread@sympatico.ca

Design by
Robbie Cooke

Printed and bound in Singapore

The publisher gratefully acknowledges the financial support for our publishing program by the Canada Council for the Arts, the Ontario Arts Council and the Government of Canada through the Book Publishing Industry Development Program.

Other Firefly books by Terence Dickinson:
Exploring the Night Sky
Exploring the Sky by Day
NightWatch
The Backyard Astronomer's Guide (with Alan Dyer)
From the Big Bang to Planet X
Extraterrestrials (with Adolf Schaller)
Other Worlds
Summer Stargazing
Splendors of the Universe (with Jack Newton)

Library and Archives of Canada Cataloguing in Publication

Dickinson, Terence
The universe and beyond / Terence Dickinson ;
foreword by Skylab astronaut Edward G. Gibson. — 4th ed.

Includes bibliographical references and index.

ISBN-13: 978-1-55297-937-2 (bound)
ISBN-10: 1-55297-937-7 (bound)
ISBN-13: 978-1-55297-901-3 (pbk.)
ISBN-10: 1-55297-901-6 (pbk.)

1. Astronomy—Popular works. I. Title.

QB44.3.D52 2004 520 C2004-903288-7

Published in Canada by
Firefly Books Ltd.
66 Leek Crescent
Richmond Hill, Ontario, L4B 1H1

Acknowledgments
Thanks to Roy Bishop (Acadia University), David Hanes (Queen's University), Robert Garrison (University of Toronto) and Alan Dyer (Calgary Science Centre), who read and commented on the manuscript for the first edition, and to Barry Estabrook, my editor, for many helpful suggestions. More thanks go to designers Ulrike Bender (first edition), Linda Menyes (second edition) and Robbie Cooke (3rd and 4th editions) and to copy editors Charlotte DuChene, Mary Patton, Christine Kulyk and Catherine DeLury. Photographs and illustrations were supplied through the generosity of many individuals and institutions listed on page 177. In this regard, Ray Villard of the Space Telescope Science Institute was especially helpful. My largest debt of gratitude goes to my wife Susan, who provided invaluable advice as copy editor, proofreader and tireless aide.

Opposite page: The Hubble Space Telescope's Advanced Camera for Surveys captured this image of the five-light-year-high Cone Nebula.

Front cover: One of the most resplendent vistas in the cosmos is the Eagle Nebula, also popularly known as the Pillars of Creation, because it contains regions of new star birth. This superb image was obtained by astronomer Jean-Charles Cuillandre using the 3.6-meter Canada-France-Hawaii Telescope atop Mauna Kea, Hawaii.

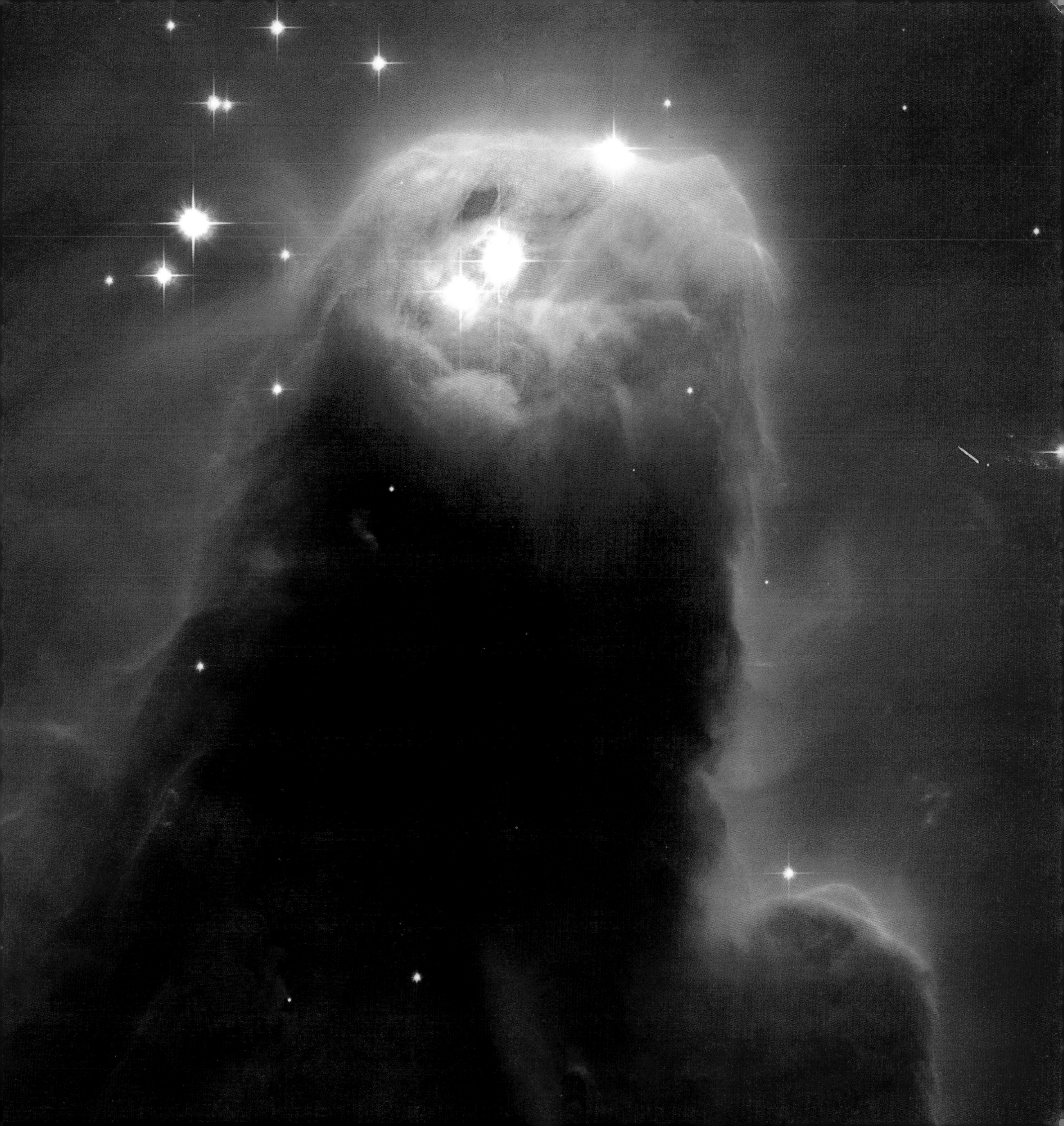

CONTENTS

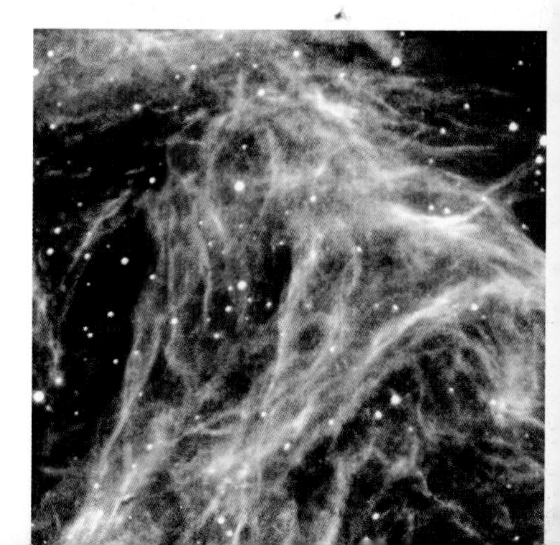

FOREWORD

The Allure of New Turf

Within each of us, there is at least the trace of a compulsion to explore. It's hardwired into our psyche. We are pulled around the next bend in the road, up into the mountains to the highest peak and across land or water to that enticing "other side." We have been to the Moon and now cast an impatient eye toward Mars, compelled to set foot on fresh turf and take full measure of it with our senses before we accept it as a real and familiar part of our world. Fortunately, two developments now allow us to satisfy this need more easily.

First, torrents of premium data from distant sources are available through new technologies, including telescope arrays and adaptive optics on the ground and the Hubble Space Telescope and other electronic eyes in space. And second, a few individuals, combining the best traits of explorers, scientists and writers, have demonstrated a genius in distilling the complexity of these streams of data into elegant and seductive prose which is both comprehensive and scientifically accurate—a genius that makes us feel as if we were actually standing right at every distant site. No one has done this better than Terence Dickinson in *The Universe and Beyond*.

To one who has been privileged to take that first micro-step off our planet—to glance back at Earth and the Sun, then gaze outward at the planets and stars—the explorations described here seem not just natural but obligatory and inevitable. We pause at our doorstep to inspect the Moon and Mars, then feel drawn, like a bear pursuing a trail of honey, to the planets and moons in the backyard of our solar system, into our stellar neighborhood and then into the stellar city we call the Milky Way Galaxy. The journey continues, with visits to nearby star cities and, finally, to the edge of all that we know.

Eventually, and quite naturally, as we continue our trek outward, we find ourselves trying to leap from the edge of our universe into the unknown beyond. Is our bubble of space just one of many universes in the cosmos, one that exploded into existence at the beginning of time like a bubble of fizz in a shaken bottle of soda pop? With deceptive ease, we move from standing with our feet planted firmly on Earth to pondering some of the grandest questions of cosmology. And all along the way, we naturally ask, Is life ubiquitous? Who else is out there?

Seeing the variety and vitality of life in every extreme on our own planet—a diverse spectrum which has spurted into existence in but a flash of cosmic time—only the dullest of minds or the grandest of egos could believe that we are alone. This is not to say, however, that we should jump to the belief that advanced life-forms (UFOs) are regularly visiting Earth just to play "peekaboo with the natives," as Dickinson reminds us.

The Universe and Beyond is a celebration of the human spirit of exploration. It is a majestic voyage to the most distant realms we can imagine, guided by a master explainer. Prepare yourself for a great adventure.

Edward G. Gibson
Astronaut, Skylab 4

Above: Astronaut Edward G. Gibson at the controls of the solar telescope aboard the Skylab space station. Gibson spent 84 days aboard Skylab in 1973-74. Facing page: Spiral galaxy NGC613, one of an estimated 100 billion galaxies in the known universe, is similar to our own Milky Way Galaxy.

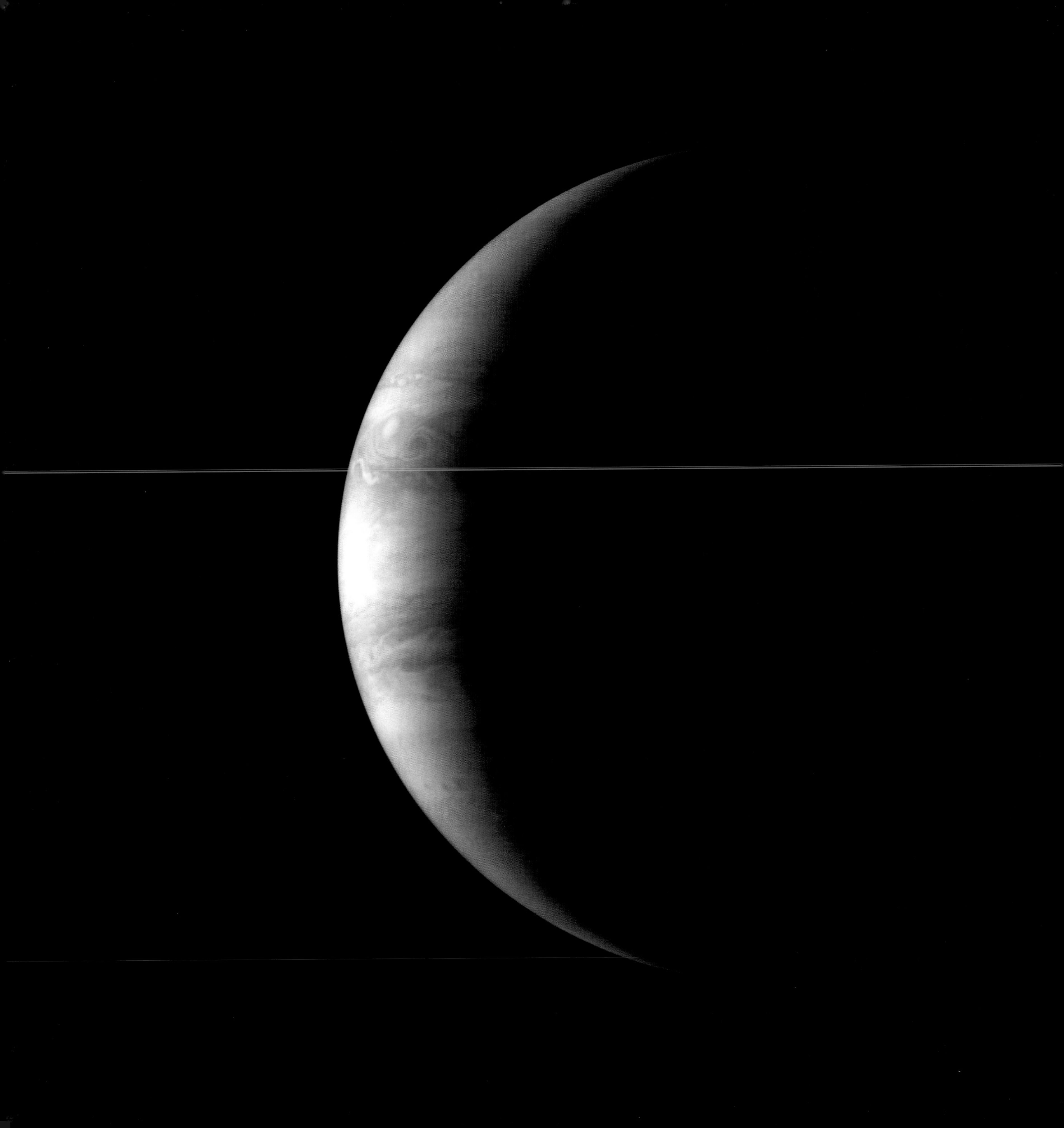

PREFACE

About This Book

If you are looking for a factual, up-to-date guide to the universe, written in accessible language and illustrated with the latest images from space probes and the largest observatories on Earth, then this is the book for you. At least, that's my not-so-humble opinion! But it's backed up by the fact that this is the Fourth Edition, a rare honor for most books.

I have attempted to cover most major topics in modern astronomy without delving into them at textbook depth, but I didn't skimp on detail. Rather, I made many hard decisions about what to put in and, especially, what to leave out. In doing so, I tried to focus on areas about which people most often ask questions. In my astronomy classes and when I am a "guest expert" on radio call-in shows and even in everyday conversation, I have noticed the same questions coming up again and again. I have included answers to as many of these as possible. But beyond that, this book is a celebration of our magnificent and, in many ways, still mysterious cosmos.

A picture may be worth a thousand words, but astronomical pictures often require some explanation before they can be fully appreciated. Therefore, many of the photograph and illustration captions in this book are fully descriptive—long and meaty to provide proper background and to augment the main text, rather than merely extract tidbits from it. Peruse the captions either as appetizers or during the reading of the main text.

The first edition of *The Universe and Beyond* was published in 1986, midway through what many regard as the golden age of astronomy—roughly the last third of the 20th century—a period during which our knowledge of the universe, both nearby and at its farthest reaches, expanded enormously. Although no one expects the pace of discovery to slow anytime soon, the beginning of the 21st century is an ideal opportunity to take an inventory of what we know and would like to learn about nature on its largest scale.

This completely revised and enlarged Fourth Edition contains significantly more illustrative material than previous editions. I am especially pleased that we were able to include four full pages of high-resolution reproductions from the 2004 Hubble Ultra Deep Field. This astonishing image is best appreciated in print, with your nose close to all those galaxies!

In a book of this size, there are inevitably some topics that we can only wave at on the way by. For those who are inspired to dig deeper, an annotated guide to further reading is offered. But I hope you will find *The Universe and Beyond* on its own a worthy celestial feast for both eye and mind.

Terence Dickinson
NightWatch Observatory
June 2004

Above: The Pleiades star cluster, also known as the Seven Sisters, in Taurus. Facing page: Crescent Jupiter, seen by the Galileo spacecraft.

A JOURNEY THROUGH TIME AND SPACE

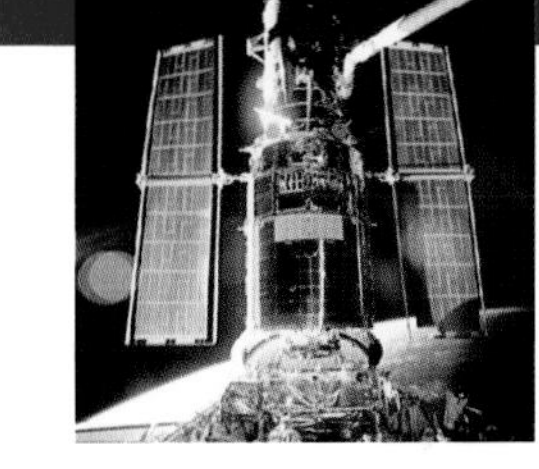

CHAPTER ONE

The spring evening is crisp and cool and pitch-black. Stars fill the sky in a glittering tapestry that goes unnoticed by the occupants of a car speeding down the rural highway, far from city lights and traffic. Dimly at first, the headlights reveal a steep hill ahead. Without losing speed, the automobile hurtles upward and reaches the crest, and for an instant, the headlamps shine into the blackness like two ghostly fingers. That instant marks freedom for at least one photon of light which avoids bumping into dust motes or being absorbed by molecules of air on its way up from the Earth's surface. Less than two seconds later, it passes the Moon. One minute after that, Earth and Moon diminish to starlike points. Within an hour, they fade into the starry backdrop. One month away from Earth, the photon is so remote that all the planets are invisible and the Sun dwindles to a star, although still far brighter than any other. In two years, the Sun is reduced to a bright but not extraordinary star.

Over the next 50 years, the Sun slowly fades until it is dimmer than the faintest stars visible to the unaided eye. Yet the sky still appears basically as it does from Earth and has the same overall proportion of bright and dim stars. But by the hundredth year of travel, a distinct thinning out of stars becomes apparent ahead. The photon is moving out of the Milky Way Galaxy.

After cruising in its arrow-straight trajectory for 2,000 years, the photon is completely outside and well above the spiral arm of the galaxy where our solar system resides. From this vantage point, only a handful of stars speckles the sky. But the view back toward the galaxy reveals an impressive panorama: the sweeping curves of the spiral arms and, beyond them, the bulging galactic nucleus. Continuing its voyage for another 22,000 years, the photon nears a mammoth swarm of stars, the Hercules cluster, a million suns congregated in a rough sphere about 75 light-years across. This is one of more than 150 globular clusters that orbit the Milky Way Galaxy like satellites.

The photon races onward, but the scenery becomes less inspiring with each millennium as the Milky Way fades to a tiny puff in the blackness. Only one additional galaxy, Andromeda, is easily visible; other galaxies appear as mere smudges. After 10 million years, both the Milky Way and Andromeda are lost to view. Millions more years will pass before chance encounters with other galaxies break the monotony of the void. And the journey has just begun.

Experiencing the Cosmos

It is no accident that the starry night stirs in us the most profound questions of origins, destinies and the ultimate meaning of it all. Humans are as much a part of the fabric of the cosmos as a Moon rock, an ice particle in the rings of Saturn or an asteroid in a galaxy a billion light-years away. We are all star-stuff, assemblages of atoms cooked in the thermonuclear fires at the hearts of stars. And before there were stars, every subatomic particle in the universe emerged from a genesis fireball—the Big Bang—trillions of times hotter than the Sun.

Our cosmic roots are revealed in another, more subtle way. Many people find that celestial objects exhibit a beauty which is difficult to quantify. Like a favorite piece of music or a painting that invites another glance, the delicate tendrils of a nebula or the

A day will come when beings shall stand upon this Earth, as one stands upon a footstool, and laugh and reach out their hands amidst the stars.

H.G. Wells
1902

Facing page: As we look past clouds of comparatively nearby stars, our gaze encounters the nearest galaxy beyond the confines of our own Milky Way Galaxy. Known as the Large Magellanic Cloud and floating 160,000 light-years from Earth, this system of several billion suns is a satellite of our Milky Way. Above: The Hubble Space Telescope has extended our reach to the edge of the universe.

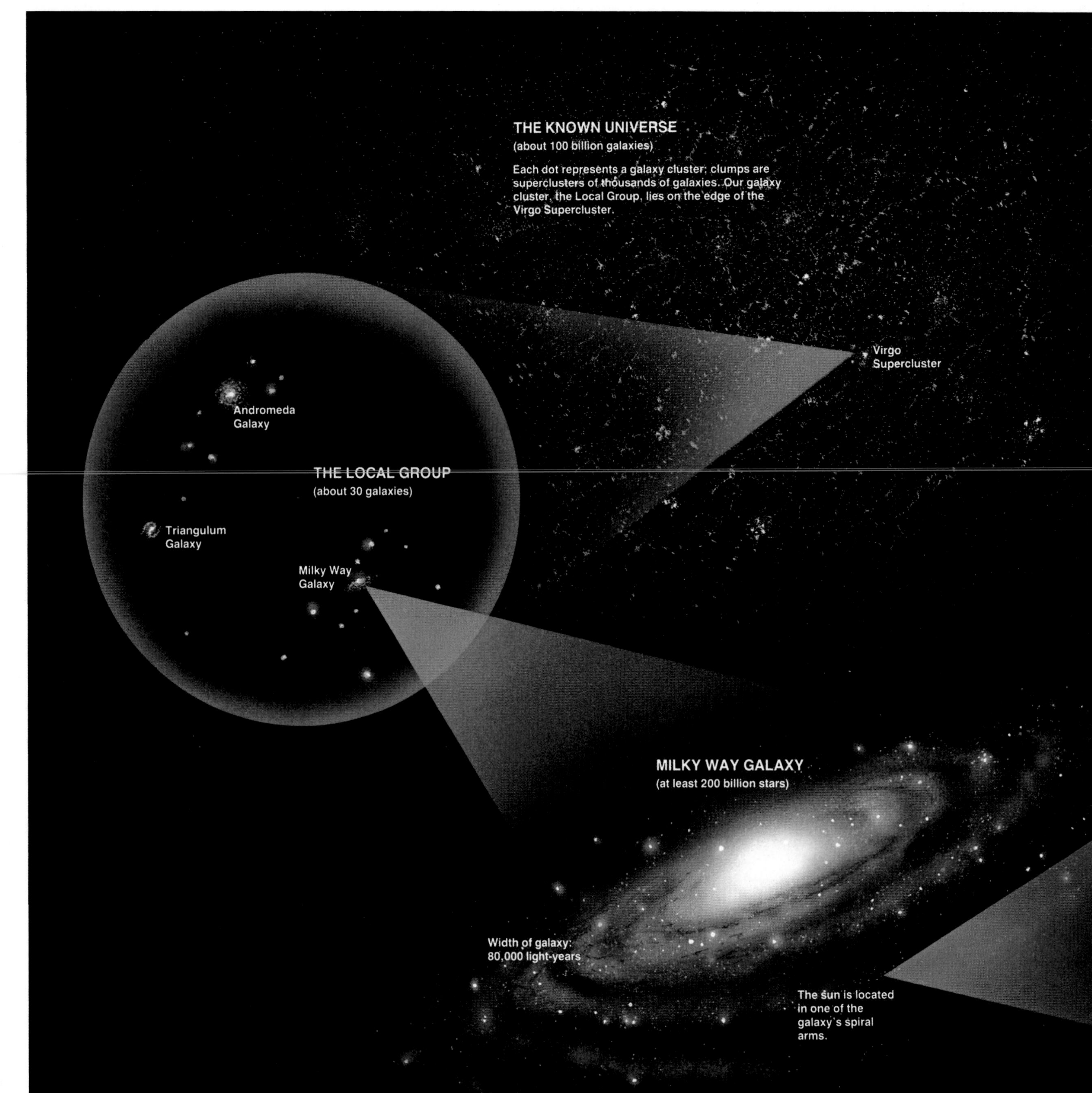
THE KNOWN UNIVERSE
(about 100 billion galaxies)
Each dot represents a galaxy cluster; clumps are superclusters of thousands of galaxies. Our galaxy cluster, the Local Group, lies on the edge of the Virgo Supercluster.
Virgo Supercluster
Andromeda Galaxy
THE LOCAL GROUP
(about 30 galaxies)
Triangulum Galaxy
Milky Way Galaxy
MILKY WAY GALAXY
(at least 200 billion stars)
Width of galaxy: 80,000 light-years
The sun is located in one of the galaxy's spiral arms.

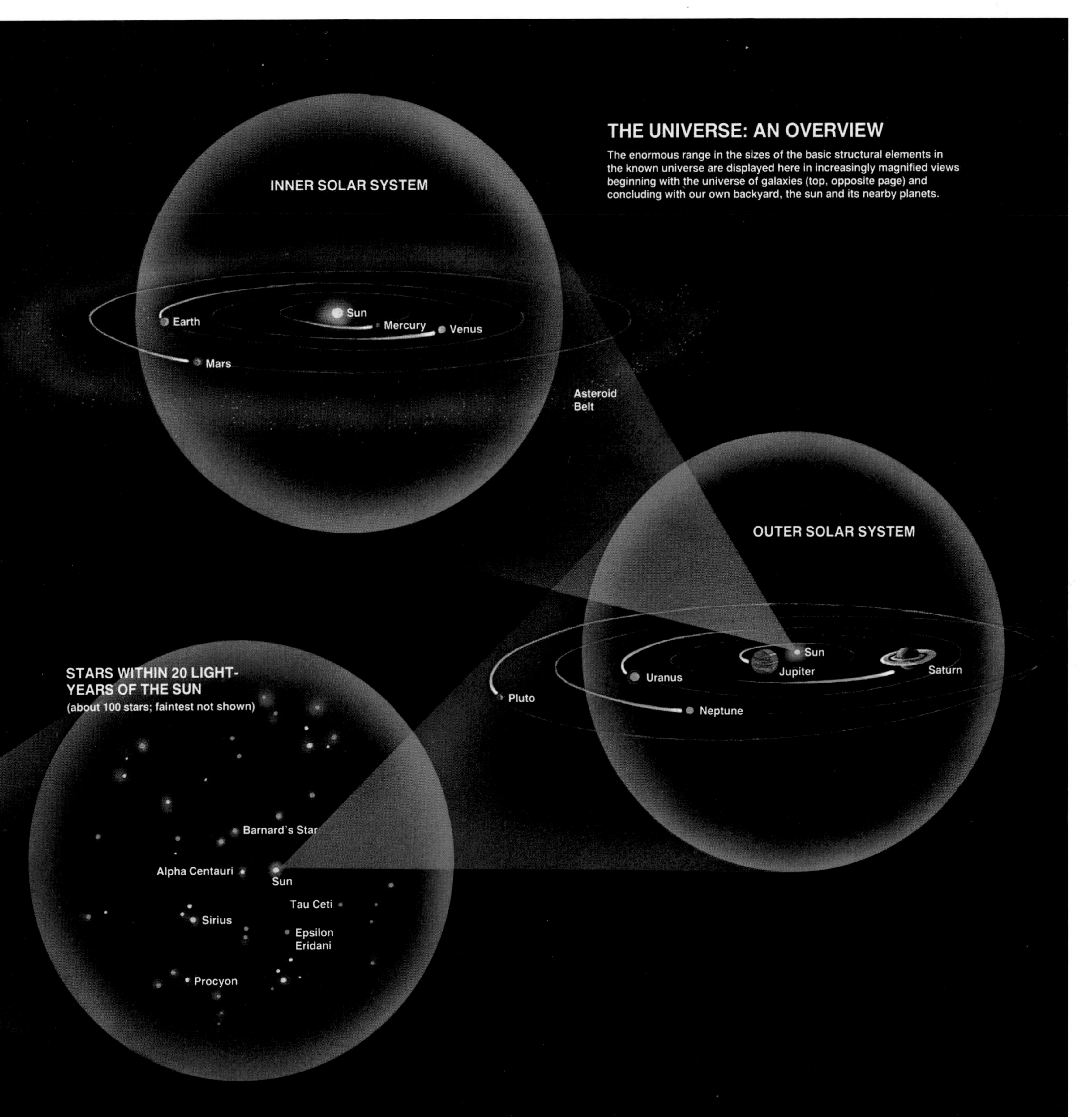

THE UNIVERSE: AN OVERVIEW

The enormous range in the sizes of the basic structural elements in the known universe are displayed here in increasingly magnified views beginning with the universe of galaxies (top, opposite page) and concluding with our own backyard, the sun and its nearby planets.

A portrait of Earth and the Moon captured in a single frame by the Galileo spacecraft reveals the stark contrast between an airless, waterless, lifeless world and our dynamic watery globe teeming with life. Now, for the first time in human existence, we are capable of seeking and examining other worlds that may nurture life as we know it.

pinwheeling arms of a spiral galaxy can be irresistible. When Canadian astronaut Marc Garneau returned from a six-day flight on the space shuttle and was asked what he found most memorable about the journey, he was emphatic: "The view of the Earth! It was incredibly beautiful. The hardest thing I had to do up there was tear myself away from the window." Fewer than 500 humans have seen our planet from space—a sapphire sphere dappled with cotton clouds and continents wearing vests of green and brown. Fewer still have ventured into the Moon's ken, and only 12 have actually walked on its dusty, cratered face.

Beyond the Moon, where no one has yet traveled, robot probes embodying electronic and computer substitutes for the human senses have allowed us to explore vicariously more than three dozen planets and moons. The red deserts of Mars, the sulfur volcanoes of Jupiter's satellite Io, the mammoth craters of Saturn's moon Mimas and the black rings of Uranus have been revealed almost as clearly as if we had been there. More distant cosmic shores, the realm of myriad stars and galaxies, remain unexplored but not unknown. Centuries of telescopic scrutiny have provided a vast inventory of facts—distances, sizes, ages—which are buttressed by theories that attempt to decipher origins and predict destinies.

What is it like out there? What varieties of worlds wheel around those distant suns? Are there other creatures in the universe that share our compulsion to know? Those questions in their broadest context are what this book addresses. It includes frequent imaginative voyages into various cosmic environments, descriptions of what it would actually be like to be there. I have left out some of the material commonly found in astronomy textbooks or encyclopedias, particularly discourses on instrumentation and how we have learned this or that fact. There are plenty of excellent books that provide such information, and a few of the best are listed at the back of this book. The approach here is more a guided tour, a celebration of cosmic wonderment.

My primary goal has been to present the various objects and elements of the universe as they would be experienced by human explorers. We stroll the sands of Mars, float among Saturn's rings, observe how one star is born and another dies and venture to planets of double suns and realms where black holes consume nearby stars or swallow whole galaxies. We travel back in time to the very origin of the universe, before anything we know today existed—to the first infinitesimal fraction of a second. I have endeavored to keep our celestial excursions rooted in current scientific knowledge. Only in the discussion of extraterrestrial life in Chapter 8 does conjecture inevitably become a major component.

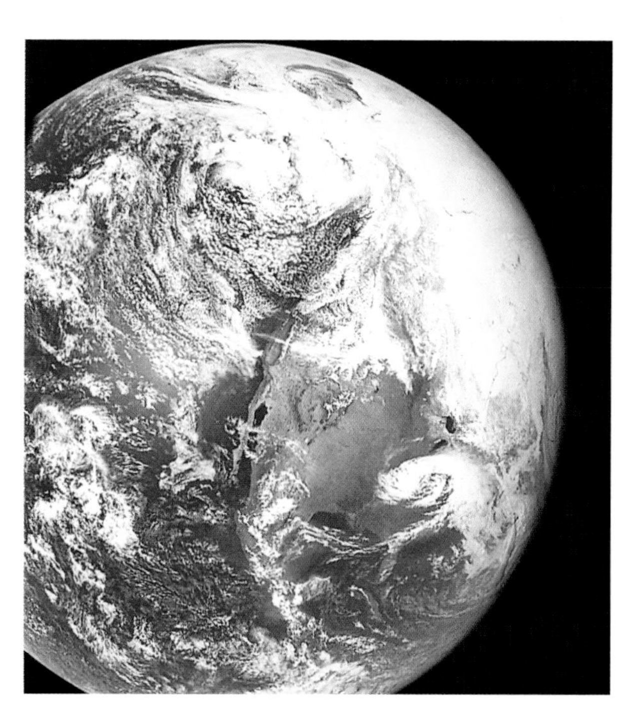

Above: The spiral galaxy M83 is one of billions of galaxies in the universe that closely resemble the Milky Way Galaxy. Left: Earth as seen by the Apollo 12 astronauts on their return flight from the Moon. The powerful impression that we live on an oasis in space was burned into the collective consciousness of humankind by global images of Earth gathered during the Moon flights.

To the Edge of the Universe

Our voyage begins with the vista of stars seen on a dark, moonless night. As each tiny spear of starlight enters our eyes, it ends a journey that began decades or centuries ago. Five of the seven stars in the Big Dipper, for example, are members of a nearby star cluster roughly 75 light-years away. Their light takes a human lifetime to reach Earth. Nestled between the constellations Perseus and Cassiopeia lies a dim blur of light that binoculars reveal as a twin cluster of stars: the Double Cluster. Their light travels for 7,000 years before it reaches us from the next spiral arm beyond the one that we occupy in the Milky Way Galaxy.

The unaided eye can bridge even greater gulfs. Overhead on late-autumn evenings, we can see the Andromeda Galaxy, a small, oval, hazy patch that is, in reality, a colossal platter-shaped stellar metropolis larger than our own Milky Way and populated by perhaps a trillion stars. The combined light of those stars is dimmed by its enormous distance from Earth: two million light-years.

When light from the Andromeda Galaxy meets a

Riding high above the atmosphere, an astronaut tests a backpack propelled by gas jets that allows untethered freedom from the space shuttle. Similar units will be used in future decades to approach comets, asteroids and the smaller moons of the solar system.

human retina and registers as an image in the brain, that particular bundle of photons terminates an uninterrupted voyage of more than two million years. One of the photons could have originated on the surface of a star like our Sun. Light from that same star might bathe a family of planets, one of which could resemble Earth. Such thoughts are central to the lure of astronomy. The possibility of other inhabited worlds in the universe is a theme I will return to as our exploration of the cosmos unfolds.

The Andromeda Galaxy is the nearest of billions of galaxies similar to the Milky Way. The Hubble Space Telescope can detect galaxies 13 billion light-years away. If the distance to the Andromeda Galaxy were reduced to the reading distance between your eyes and this book, the most remote galaxies known would be nearly two kilometers away. Their images in a photograph (or, more likely, computer-enhanced images gathered by an electronic detector more sensitive than photographic emulsion) are ghostly smudges. Yet each wisp represents the combined light from billions of suns that has been on its way to

Above: The Trifid Nebula in the next spiral arm of the Milky Way Galaxy is one of many centers of star birth in our galaxy. Our Sun and its planets emerged from a similar nebula about 4.6 billion years ago. Left: A swarm of a million stars, the globular star cluster M13 is about 22,000 light-years from Earth. M13 and a family of more than 150 other globular clusters orbit around the central gravitational hub of the Milky Way Galaxy.

Earth since before our planet or the Sun even existed.

Thousands of galaxies have been studied and cataloged, and millions have been photographed dusting the blackness of space, like snowflakes in a cosmic blizzard frozen in time. Billions of star cities are thought to exist, but the universe does not go on forever. It has finite boundaries in both space and time—about 14 billion years' worth—before which there was nothing. As we shall see, one of the great triumphs (or perhaps conceits) of modern astronomy is its ability to offer at least an approximate explanation for the universe's origin and evolution from time zero to today.

Billions and Billions

Like nothing else in human existence, astronomy encompasses distances, ages and sizes so enormous that it is almost impossible to form a mental image of any of them. The numbers alone are enough to overwhelm the imagination. How big is a million or a billion? A million sheets of paper about the thickness of

At a distance of 2.4 million light-years, the Andromeda Galaxy, right, is the nearest major galaxy to the Milky Way. Flanked by two smaller satellite galaxies, the Andromeda Galaxy is slightly larger than our galaxy and is the most distant object visible to the naked eye. At the opposite extreme is the view on the facing page. Known as the Hubble Ultra Deep Field, it is a 275-hour Hubble Space Telescope exposure designed to penetrate deeper into the universe than any previous images. Everything in the image is a galaxy. The faintest galaxies visible are more than 12 billion light-years from Earth; the prominent yellow spiral galaxy is about 800 million light-years distant. To photograph the entire sky at this level of detail would require one million years of continuous imaging with the Hubble telescope. This image is about one-sixth of the entire Ultra Deep Field frame. The remainder of this remarkable image is displayed on pages 118 and 132-33.

the pages of this book would form a tower as high as a 30-story building, but a billion-page stack (1,000 million) would soar more than 10 times higher than Mount Everest. And a trillion pages? (One trillion is a million million, or 1,000 billion.) The pile would reach more than one-quarter of the way to the Moon.

Another way to think of it is that a million seconds is 12 days, but a billion seconds is more than 31 years, and a trillion seconds is 300 centuries, longer than the history of civilization. Yet even with these analogies, it is often less than edifying to rely solely on huge numbers. Sometimes a better perspective can be gained by examining a model of the cosmos.

Let's start close to home. If the Sun is reduced to the size of a ping-pong ball, Earth becomes a mote of dust 2.5 meters from it, with a smaller speck, the Moon, nestled beside it six millimeters away. Jupiter is a pea 12 meters from the Sun. Pluto, the outermost planet in the solar system, the Sun's family, is a piece of dust 100 meters from the Sun. Comets are atom-sized particles, invisible to a microscope, extending in a cloud up to several dozen kilometers from the Sun. Although there are trillions of comets, the vast volume of space they occupy keeps them, on average, several strides apart.

Alpha Centauri, the nearest star (a triple-star system), consists of two walnuts and a pea 700 kilometers away. Even if the universe were shrunk to this microscopic scale, it would be inconvenient to hike to the nearest star. If the model were encompassed in a volume of space the size of Earth, the vast hollow globe would contain only 800 stars, represented by walnuts, cherries, oranges, and so on. The billions of other stars in the galaxy would range well outside the Earth-sized region. Clearly, the universe is mostly empty space.

To picture a larger portion of the universe, a further reduction of the model is necessary. The Earth's orbit is now the size of a dime. Most of the stars are dust-sized, and gigantic Betelgeuse, the largest star known, is a golf ball. The average distance between stars has shriveled to one kilometer, and the entire Milky Way Galaxy is as wide as the real diameter of Earth. Our neighboring galaxy, Andromeda, is a little over half the Earth-Moon distance from us. If, like gods, we could stroll through our miniature universe, we would move from one star to another in a few minutes. But the trek between galaxies would still take more than a lifetime.

Let us make one final reduction: The Milky Way Galaxy now becomes small enough to hold in your hand, a delicate disk of subatomic-sized stars. The Andromeda Galaxy, slightly larger, is not quite within reach at its scaled-down distance of two meters. The universe of galaxies can be perceived in all directions, ranging from basketball-sized giant ellipticals to dwarf galaxies no larger than the head of a pin. Yet the immensity of the cosmos still defies the senses. There are so many galaxies in the universe and they fill a volume of space so vast that even though they are just a few steps apart, it would take centuries simply to walk past all these tiny islands of stars. And although we think it impossible to know what lies beyond our universe, there is no reason to presume that ours is the only one. The bubble of space we call the universe could be but one in an infinity of universes that constitute the true cosmos.

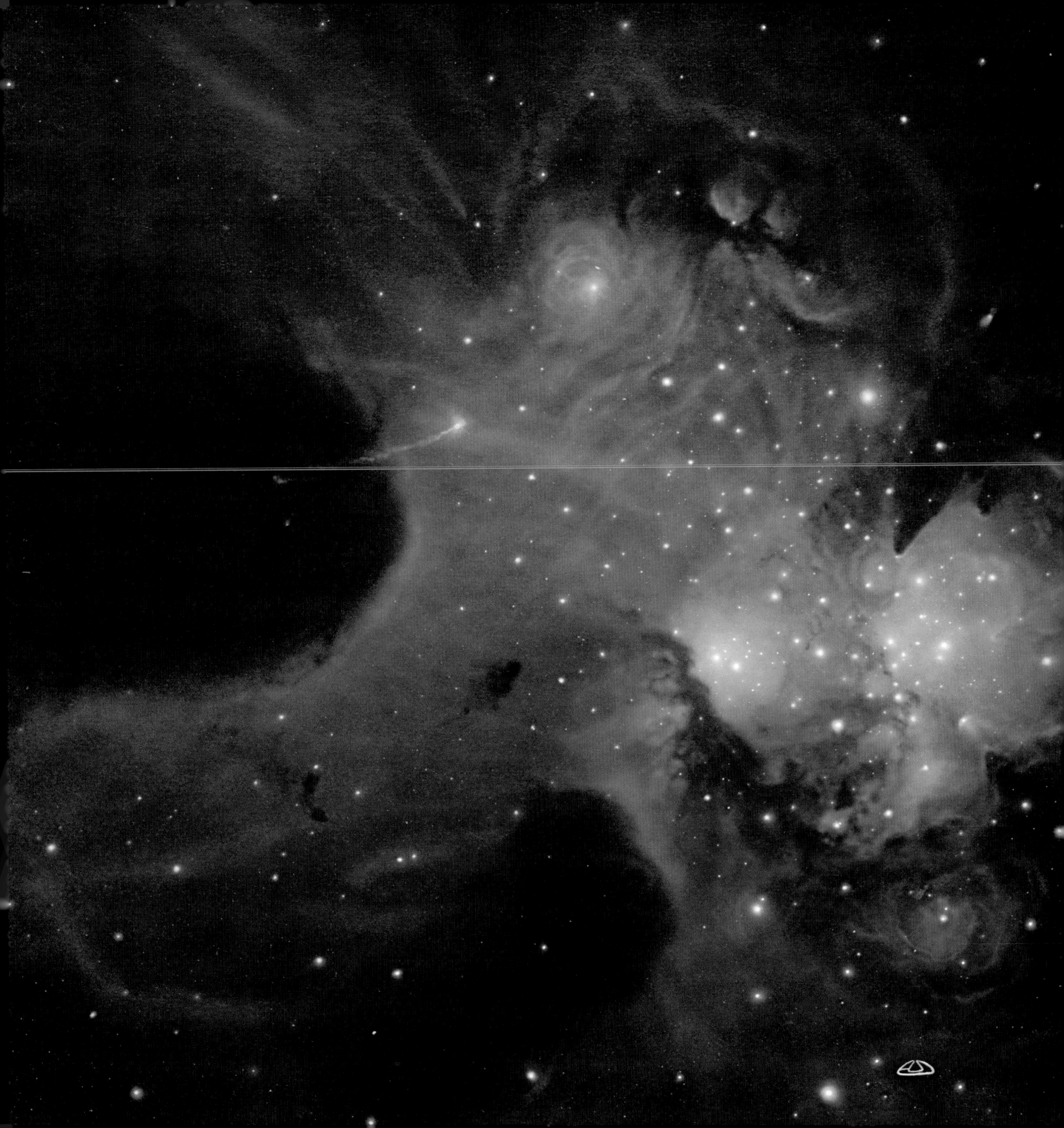

NEARBY WORLDS

CHAPTER TWO

Five billion years ago, a dark, formless nebula—a hundred million cubic light-years of diffuse gas and dust—floated within the Milky Way Galaxy's graceful spiral arms. The vast cloud was one of thousands that populated the galaxy. While some of these dispersed, others would undergo an impressive transformation, a metamorphosis that was to be the fate of this particular cloud.

Barely perceptible at first, a pocket of the cloud began to contract, perhaps recoiling from the shock wave produced by a nearby supernova—the violent death of a massive star—or possibly as the result of a merging of several clouds. Whatever triggered it, the collapsing region escalated in density when the contraction started; atoms once comfortably separated were jostled more vigorously, generating heat. In less than 100,000 years, nebular material at least a million times the Earth's mass had collected in a zone several times wider than the present orbit of Pluto, the most remote of the Sun's known planets. At its heart, a seething ball of hot gas, stoked by the crushing pressure of infalling matter, reached the ignition point for nuclear reactions: The Sun was born.

Meanwhile, the cloud material in the zone surrounding the primal Sun had swirled into a disk, like a miniature spiral galaxy, called the solar nebula. The disk shape emerged because whatever internal motion the original cloud had was amplified during the contraction, just as a figure skater accelerates a spin by drawing in her arms. The process pulled much of the infalling material into the disk.

The original cloud contained abundant quantities of hydrogen and helium, lots of oxygen, nitrogen, carbon and neon, moderate quantities of iron, silicon, magnesium and sulfur and lesser amounts of all the other elements. Much of the material had been through this same routine before in previous generations of stars, which had returned it to space when they ended their lives in supernova explosions or other, less flamboyant forms of stellar termination.

Now swimming at closer range than in the near-vacuum conditions of the initial cloud, atoms and molecules within the nebula began to combine into larger particles in somewhat the same way that ice crystals—the precursors of rain and snow—form in the upper atmosphere when air becomes saturated with water molecules. Over the next few million years, the particles combined like adhering snowflakes, eventually accumulating into larger and larger bodies called planetesimals.

The planetesimals, in turn, collected into planets, moons, asteroids and comets. The composition of these bodies varied depending on their distance from the Sun, producing a bull's-eye pattern. Closer in, the Sun's radiation efficiently swept away most of the light gases, leaving rocks and metals (Mercury, Venus, Earth, the Moon and Mars formed here). Farther out, carbon-rich composites similar to charcoal and dirt dominated, with rock and metal mixed in (majority of the asteroids). More distant still, materials rich in carbon-nitrogen-water compounds were common (outer asteroids and small moons of Jupiter). The sector of the solar nebula beyond this was cool enough for water-ice particles to remain, along with hydrogen and helium gas (Jupiter and Saturn and their major moons). The most remote realm was frigid enough for ammonia, methane and carbon-monoxide snow (Uranus and beyond).

Mankind is made of star-stuff, ruled by universal laws.

Harlow Shapley
1962

The birth of the Sun and its family of planets took place 4.6 billion years ago in a dark corner of a collapsing cloud of gas and dust—the genesis nebula for hundreds of new stars. The incipient Sun, a dull red ember in the black mass that extends from the center to the left in this rendering, is glowing from the heat of compression as the cloud material collapses under its own weight. A few dozen stars in another sector of the cloud at right have just come to life through ignition of their internal thermonuclear fires. Radiation pressure from the newborn stars peels back the nebula to expose the cloud's interior.

The crater-battered face of the Moon, above, is a cosmic museum dating back to an era of violent collisions that marked the first half-billion years of the solar system's existence. A young Earth, right, was bombarded just as heavily as was the Moon, but aeons of weathering, erosion and mountain building have erased almost all traces of the primordial concussions.

We are most familiar with the inner zone, the region of rock and metal, because we are in it, along with the Moon, our companion in space, and the neighboring worlds Mars, Venus and Mercury. Robots have probed and analyzed the surfaces of Mars and Venus and have gathered high-resolution pictures of Mercury from a flyby mission. But it is the Earth-Moon system we know best.

Earth and Moon: A Gravitational Embrace

Earth and the Moon are a binary planet system. If the Moon were circling the Sun in its own orbit, say, between Earth and Mars, astronomers would not hesitate to classify it as a planet in its own right. Its diameter is two-thirds that of Mercury, significantly larger than Pluto, the smallest planet, and one-quarter the Earth's diameter. Yet there it is, obediently swinging around Earth. None of the other terrestrial planets has anything resembling our Moon. Venus and Mercury have no moons, and Mars' two Manhattan-sized companions are microscopic on a cosmic scale. Even when compared with the moons of the giant outer planets, our satellite is more than respectable. Jupiter's Ganymede and Saturn's Titan are the size of Mercury, but their parent planets are colossal relative to Earth. Pluto, which is so small that astronomers continue to debate its planet status, is endowed with a moon half its diameter. But Pluto is different in so many ways from all the other planets that it offers a poor basis for comparison.

Just how our Moon was created and came to be orbiting Earth was a mystery until the 1980s. For more than a century, astronomers had debated the merits of three scenarios: the adopted-cousin theory (the Moon was a small planet gravitationally captured by Earth); the sister theory (Earth and the Moon were born as a double planet); and the daughter theory (the Moon fissioned from a rapidly spinning primordial Earth). Yet studies of the nearly one ton of lunar material returned by the Apollo astronauts failed to support any of these theories. Instead, a fourth hypothesis emerged and is now widely accepted.

The new scenario—we'll call it the chip-off-the-old-block theory—is a product of modern computer simulations of the formation of the solar system. The simulations suggest that 10 million years after the solar nebula initially evolved, the material in the region where Mercury, Venus, Earth and Mars were emerging had built up thousands of mountain-sized

planetesimals. These, in turn, collected into perhaps a dozen bodies in the Mercury-Mars zone. Earth might have had four or five neighbors up to three times the mass of Mars, all closer than Mars is now. Then: Bang! One of them smashed into the nascent Earth. The heat generated by a collision with something that big would have melted the surface of our young planet. Debris from both Earth and the impacting body vaporized and splashed into nearby space. Some of it fell back to Earth, but a portion lingered in orbit around Earth, eventually coalescing into the Moon.

That, in brief, is the chip-off-the-old-block theory. Astronomers have accepted it because it's the only Moon-origin theory that fits with the Moon samples, which revealed that Moon material is substantially different from Earth rock—so much so that the sister and daughter theories became untenable. The Moon contains very little iron and lacks such volatiles (more easily vaporized substances) as water, chlorine and potassium, which indicates that the Moon was at one time heated to incandescence. Melting caused by the giant primordial impact would account for the

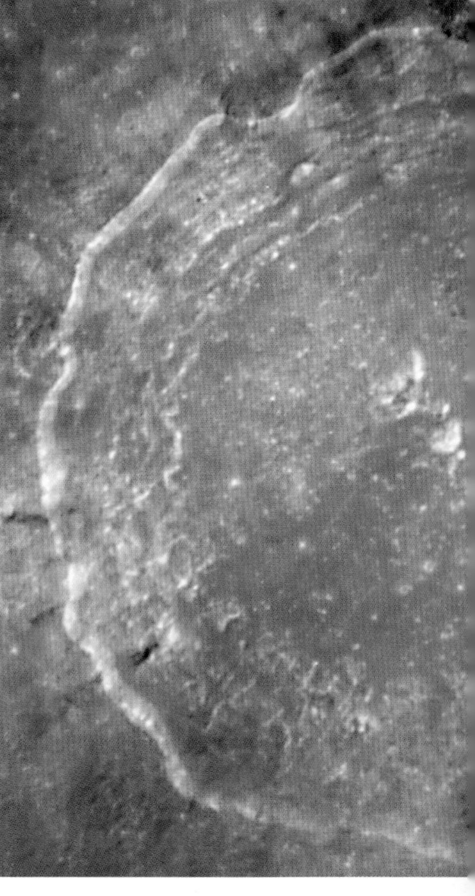

Left: Pounded by a nearly continuous barrage of rubble left over from the formation of the solar system, primordial Earth glows from the impact-generated heat. Nearby, the newly minted Moon swings through rings of debris remaining from its birth. In this early era, just 10 million years after the solar system's creation, the Moon had a much smaller orbit than the one it plies today. The Moon has gradually receded from Earth ever since. Above: Part of the lunar crater Copernicus, photographed by the Hubble Space Telescope.

shortage of volatiles, and the iron deficiency makes sense because the splashed-out material that formed the Moon would have come from the surface of Earth and of the rogue planet, rather than from their more iron-rich cores. The composition of the entire Moon seems to be more like the Earth's mantle and crust than its interior. Final confirmation came in 1999 when results from the Moon-orbiting Prospector mission showed that the Moon has a tiny iron core—proportionately much smaller than the cores of the inner planets—which is just what would be expected from a collisional origin.

The chip-off-the-old-block theory also explains the uniqueness of the Earth-Moon system among the inner planets. A big moon would result only from a collision at just the right angle early in the solar system's history. Most collision trajectories would produce a smaller moon (or moons) or none at all. Thus it is not surprising that only one oversized moon exists in the inner solar system. (The adopted-cousin theory has consistently failed the most sophisticated computer models and has been shelved by researchers.)

Although the Moon is comparatively close in size to Earth, as satellites go, the surface conditions on the two worlds are completely different. Those now historic images of spacesuited astronauts bounding over dusty lunar plains tell much of the story. The Moon is the ultimate desert: no air, no water. A grilled wasteland during the two-week lunar day, it is a freezer, almost as cold as deep space, for the two-week night. It all stems from having a relatively small mass, with its consequently weaker gravity (one-sixth that of Earth). Any air or water the Moon once had could not have lasted for more than a few million years. The simple heat of sunlight would give air and water molecules sufficient energy of motion to escape the weak lunar gravity and become dispersed into space.

Smaller mass also means reduced pressure of compaction on the interior of the Moon, compared with Earth. Planet-wide modifications caused by volcanoes and seismic activity are almost nonexistent on the Moon. It is geologically quiet: Seismometers left by the Apollo astronauts easily registered vibrations when a baseball-sized meteorite struck the lunar

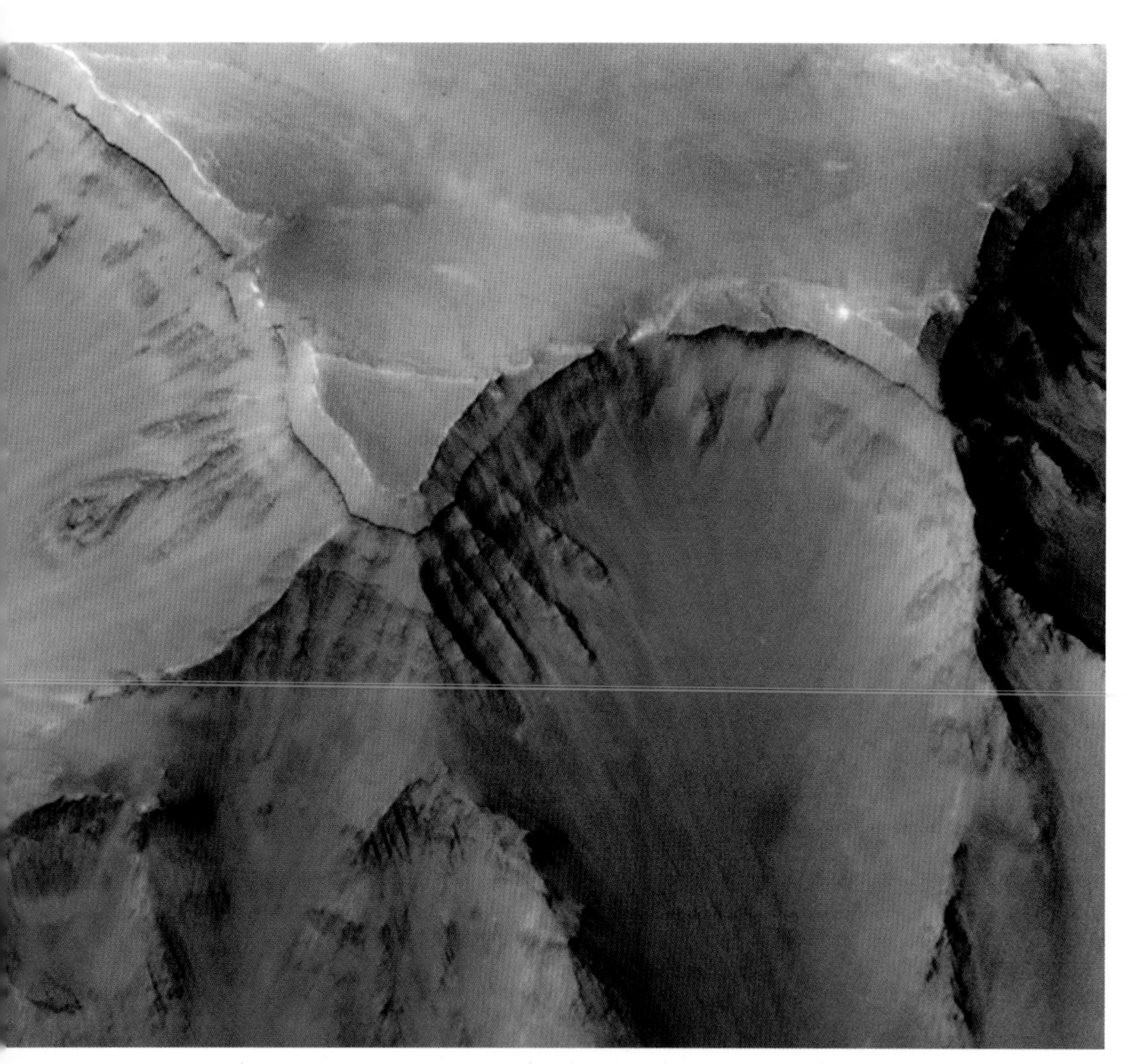

surface on the opposite side of the Moon. Major surface-shaping forces on Earth, such as plate tectonics (continental drift) and midocean ridges, are completely absent on the Moon. Its surface is a museum that records not internal but external processes. The craters are evidence of impacts from without—yawning bowls violently excavated during the solar system's exuberant youth.

Earth suffered the same bombardment. Indeed, as mentioned above, a collision that almost destroyed Earth appears to have been the event which created the Moon. Primitive Earth, cratered and without today's atmosphere and oceans, probably resembled the Moon. Over time, steam and gas from volcanoes could have supplied our planet with water and air. More water would have arrived via impacting comets, which are largely composed of water ice. The atmosphere has evolved since then, but little escapes the Earth's gravitational grip. Environmental forces helped erase evidence of craters through weathering (wind, rain, rivers, glaciers), while the Earth's internal engine restructured the surface through plate tectonics, mountain building and volcanic activity. That plus the Earth's temperature, which is primarily dictated by its distance from the Sun, have made Earth the only world in the solar system that has sustained large amounts of liquid water on its surface for billions of years. That, in turn, provided the crucible for life, which has existed on this planet for at least 3.6 billion years.

Above: The edge of Valles Marineris, an immense Martian canyon more than six kilometers deep, is a remarkably Earthlike landscape. Other sectors of Mars look more like the Moon (see page 34). Right: Three Martian volcanoes stand in stark relief against a flat plain that was formed by lava flows during the past one to two billion years. Note the wispy water-ice clouds near the tallest of the volcanic trio. Both images on this page are from the Mars Global Surveyor mission.

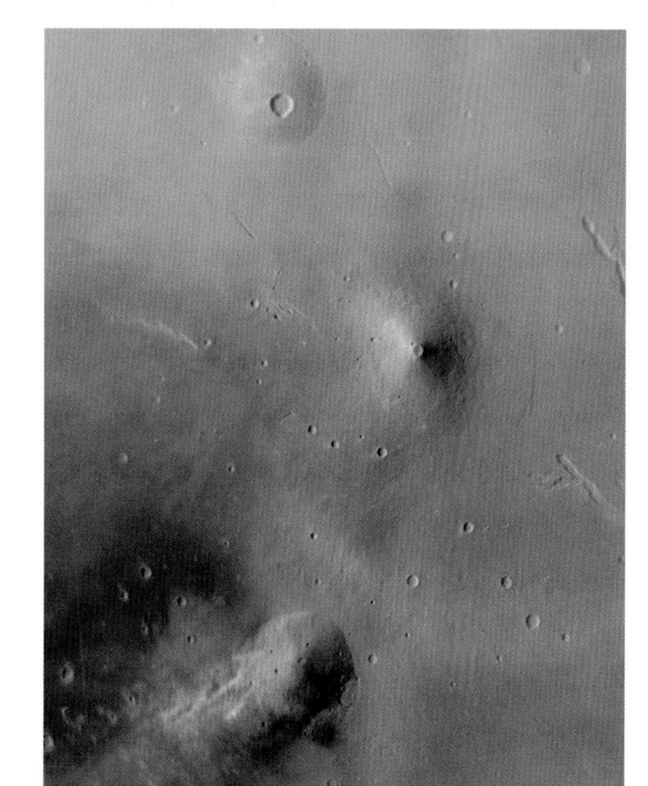

Mars: A Nice Place to Visit, but...

Mars is not an unfriendly place. In many ways, it is the most Earthlike planet we know about. A balmy day on Mars would resemble a summer afternoon in Antarctica. But an explorer would need a spacesuit, since the air is not breathable. It is mostly carbon dioxide, and there is not much of it—its density is less than 1 percent that of the Earth's atmosphere.

The main problem, however, is the cold. At noon on the hottest day of the year near the equator, the top few millimeters of the soil might warm to room temperature for a few hours; but the thin atmosphere never gets that warm. When the air temperature even approaches the freezing point of water, it is a Martian heat wave. In any case, the air pressure is too low for water to exist as a liquid. By sunset, both air and ground are as frigid as a winter day on Baffin Island. At midnight, any place in Antarctica would be warmer than it is in Mars' equatorial zone. In the Martian polar regions in winter, the bottom really falls out,

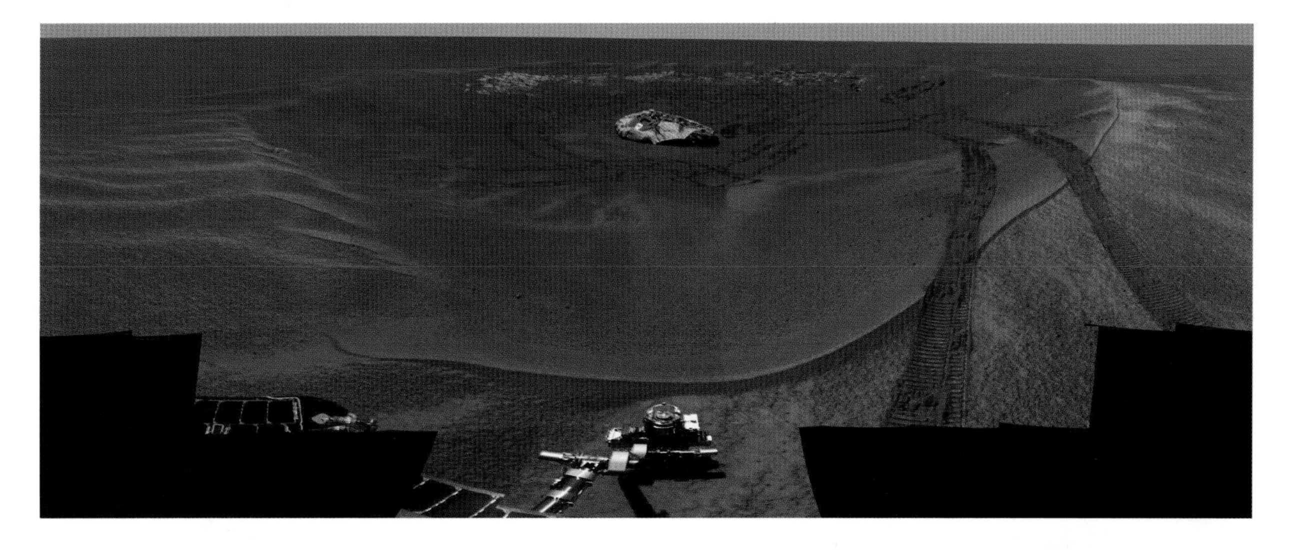

Above: This illustration shows one of the identical rovers, Spirit and Opportunity, that landed on opposite sides of Mars in 2004.
Left: By pure luck, the Opportunity rover touched down near the center of a small Martian crater. Opportunity departed the landing platform (seen at center) but remained within the crater examining rocks for six weeks. Wheel tracks show this activity and the rover's eventual climb out over the rim. A total of 558 individual images were digitally assembled to produce this panorama. Opportunity's boom-mounted camera and geology kit are visible in the foreground.

with daily high temperatures in the vicinity of minus 110 degrees C. That is cold enough for the Martian air to freeze—and it does, producing a frosty blanket of dry ice (frozen CO_2) over the winter pole.

Subzero temperatures will not stop future Mars explorers, who will be outfitted in updated versions of the fully heated and air-conditioned spacesuits that are available today. Getting humans to Mars and back is technologically feasible, given the commitment and the funds. It would take about two years for the round-trip, so one of the major hurdles is designing life-support systems for such long-duration voyages away from Earth-based supply ships.

In 2003, the Bush Administration in the United States proposed an initiative to resume human exploration of the Moon and to proceed with human expeditions to Mars—though, crucially, no time line was mentioned. Following the announcement, several experts agreed that spacecraft carrying crews to Mars are possible but estimated that the cost would be in the hundreds of billions of dollars compared with about one billion dollars for a typical robotic mission, such as the rovers Spirit and Opportunity that explored Mars in 2004. Humans will undoubtedly travel to Mars, but my guess is that the mission will not be launched before the middle of the 21st century.

A Stroll on Mars

On a summer afternoon in the Martian Amazonis desert (a possible landing site), an explorer would be struck by the color—or lack of it. Everything is shades of orange, ranging from the bright salmon-hued sky to the rusty orange dust that covers the surface. The thin atmosphere filters little sunlight, so daylight is almost as strong as it is on Earth, although the Sun appears half the size because of its increased distance. On the horizon, enormous sand dunes rise and fall in a procession of frozen waves. The ground is firm in some places, powdery in others and littered with pebbles that often more closely resemble cinders than smooth or granitic Earth rocks. There are no mountains or cliffs, at least not in this region. Certain sectors of Mars have extremely rugged terrain—canyons, volcanic flows, crater walls—while huge tracts are relatively flat.

A cloud of dust stirred up by the flick of an astronaut's boot would be quickly dispersed by the sharp breeze. Summer is the windy season on Mars. Gusts can exceed 300 kilometers per hour, but with no exposed skin, a spacesuited explorer would not feel them. Nor would the astronaut be buffeted by the blasts, since the low-density air has far less force than the winds on Earth. However, fine dust particles from the orange desert are swept into the atmosphere, producing the peach-colored sky. The strong seasonal winds kick up so much dust that the sky is never sufficiently free of it to expose the true hue of the Martian air—a deep bluish purple.

The color of the Martian sky came as a total surprise to scientists when the first pictures of the surface of Mars were returned from the U.S. Viking 1 spacecraft that landed on the planet in July 1976. In the rush to get images out to the media, the initial digitally encoded color pictures received by Mission Control at the Jet Propulsion Laboratory in Pasadena, California, were processed to produce a deep blue sky, which is what everyone had expected. This gave

A panoramic view of Mars taken in 1997 by one of the cameras atop the Pathfinder lander shows a Martian landscape that must be typical of vast tracts of the desert planet. The rumpled, cream-colored material in the foreground is the deflated air bags used to cushion Pathfinder's landing on Mars. Wheel tracks made by the rover Sojourner mark its path from the end of the ramp at right to a desk-sized boulder.

the surface material a muddy greenish brown cast.

But that was not the true Mars. By the following day, the researchers had had more time to examine the color-calibration data provided by the spacecraft, which revealed that the sky was, in fact, pinkish orange. When the photographs were reprocessed, the Martian surface took on its now familiar rusty orange color. The incident exemplifies Earthlings' generally unconscious but pervasive tendency to transport terrestrial biases to other worlds.

True, Mars has some uncanny similarities to Earth. Its rotation period, its day, is 24.6 hours. The planet's axis, tipped at almost the same angle as the Earth's, produces seasons analogous to ours, although because of the planet's larger orbit, a Martian year is 1.9 Earth years. But Mars was born both too small and too far from the Sun. A larger planet with a correspondingly greater gravitational attraction might have been able to retain a denser atmosphere, which would absorb more heat from the Sun, raising the surface temperature and allowing liquid water to exist.

At one time, Mars was probably much more like Earth than it is today. More than one hundred channels that look like dry riverbeds extend over hundreds of kilometers of the planet's surface and were likely carved by flowing water. But if water ever existed as lakes or oceans sometime in the past, it was billions of years ago. The channels themselves are old and parched, unnourished for aeons. Some of the original water is almost certainly still on Mars, but it must be in the form of ice—permafrost—frozen like cement below the surface, as it is in the Arctic regions on Earth. Another ice-storage locker is in the polar caps, which may be up to a kilometer thick (the frozen carbon dioxide is simply a frosting that covers the polar regions and extends beyond the water-ice polar cap during winter at each pole).

The channels indicate episodic and perhaps catastrophic releases of water on Mars, possibly due to heat from the interior melting a region of permafrost and releasing a cargo of water that flooded the surface. Judging from the appearance of the channels, it was not long before the water evaporated into space or seeped underground and refroze. Under the condi-

Left: As Mars rotated through its 24.6-hour day, the Mars Global Surveyor spacecraft captured these six views displaying the weather and surface features on our neighbor world: white polar caps, dust storms and the light- and dark-colored deserts. The Martian atmosphere is thin (less than 1 percent the density of the Earth's atmosphere), but high winds occasionally kick up dust storms that blanket most of the planet (note major storm at bottom of middle view, lower row).

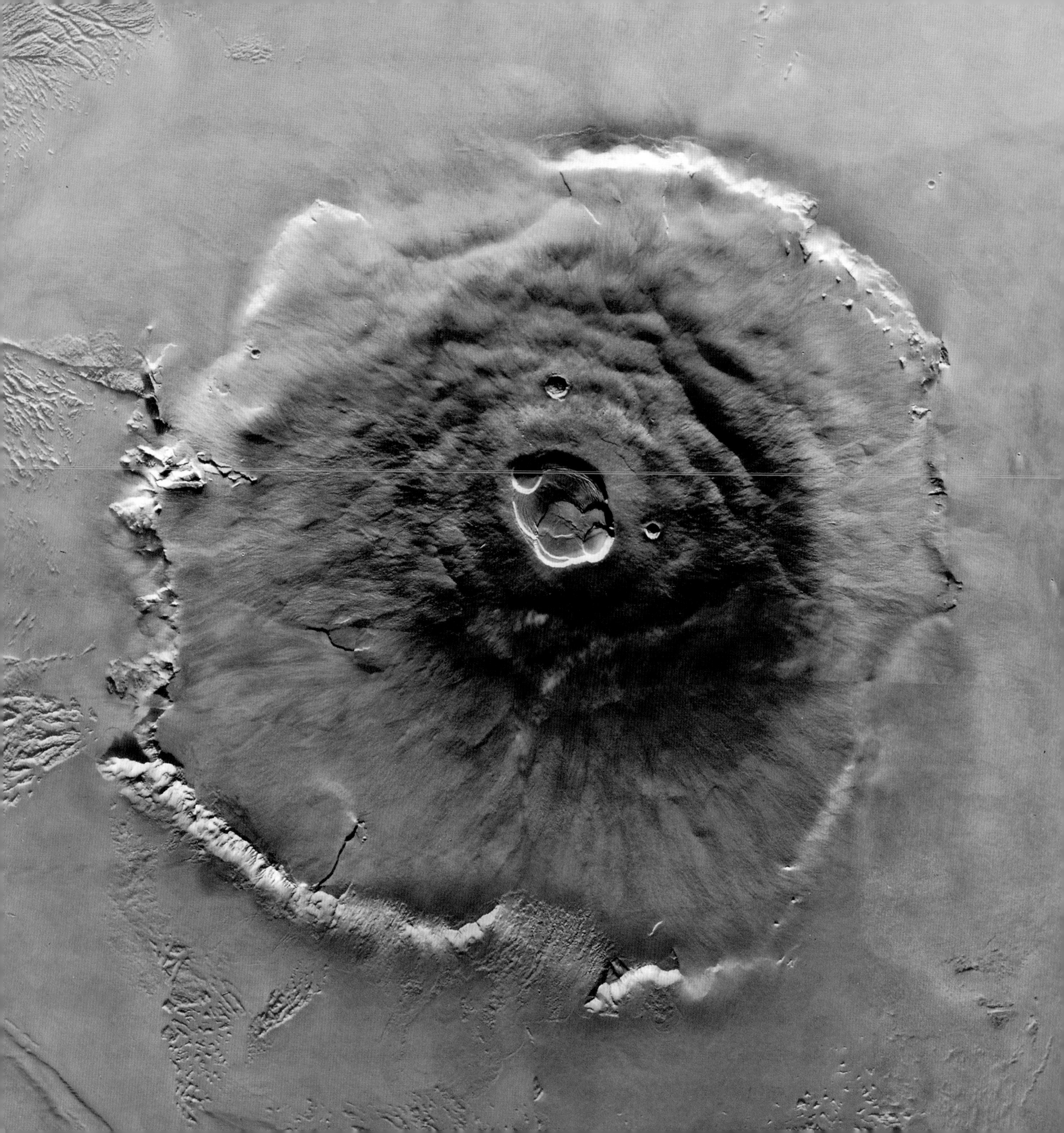

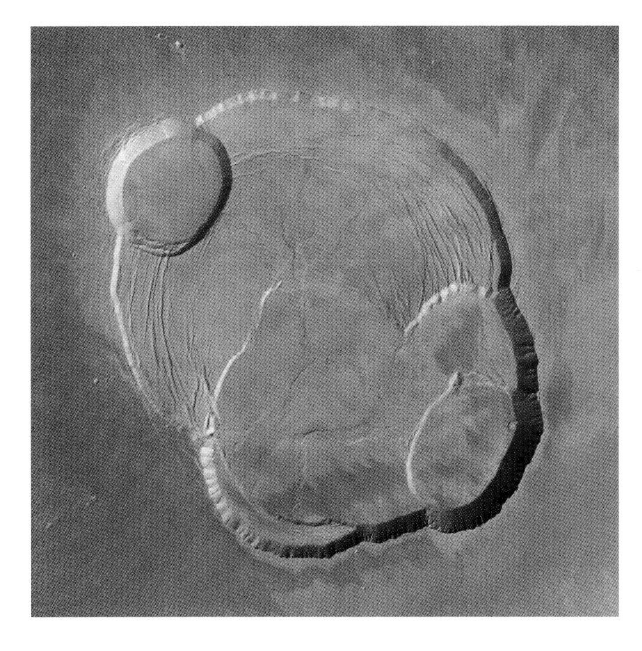

tions that exist today, a bowl of water placed on the surface of Mars would vaporize so quickly, it would literally explode. The atmosphere is starved for moisture. Yet some planetary scientists predict that if the red planet could somehow be warmed up, the permafrost melted and the atmosphere made a little denser, there could be large lakes, if not oceans, on Mars.

The idea that Mars has had liquid water on its surface in the past graduated from being a theory to a near certainty thanks to the 2004 rover Opportunity. When Opportunity's cameras first scanned its surroundings, mission controllers back on Earth realized that the spacecraft had landed in a 22-meter-wide crater with an outcrop of sedimentary rock just a few meters away (see photo, page 25).

Sedimentary rock—the type that forms at the bottoms of lakes, ponds and rivers—had not been seen in any previous Mars exploration. Further, the 2004 rovers, Spirit and Opportunity, were the first to be equipped with a full range of geological tools and experiments for grinding and analyzing the Martian rocks. Subsequent analysis of the sedimentary rock revealed fossilized ripples just like those seen on lake and river bottoms. In addition, Opportunity found small BB-sized "blueberry" rocks everywhere. Geologists recognized these as the type that form when groundwater percolates through rocks.

Within two months of the rovers' landings, scientists were satisfied that they had more than enough evidence to go public. At a news conference in April 2004, mission scientist Steve Squyres announced, "Liquid water once flowed through these rocks." He went on to say, "This was water you could swim in, not just groundwater."

Surface water indicates that conditions on Mars were once significantly warmer than they are today—possibly for hundreds of millions of years. The atmosphere had to be denser and more Earthlike for water to exist as a liquid. The question now is, How long did that period last?

To find out, at least one new robot spacecraft will be heading to Mars every two years until at least 2015. (A launch "window" opens every 26 months, when Earth and Mars align.) The Mars Reconnais-

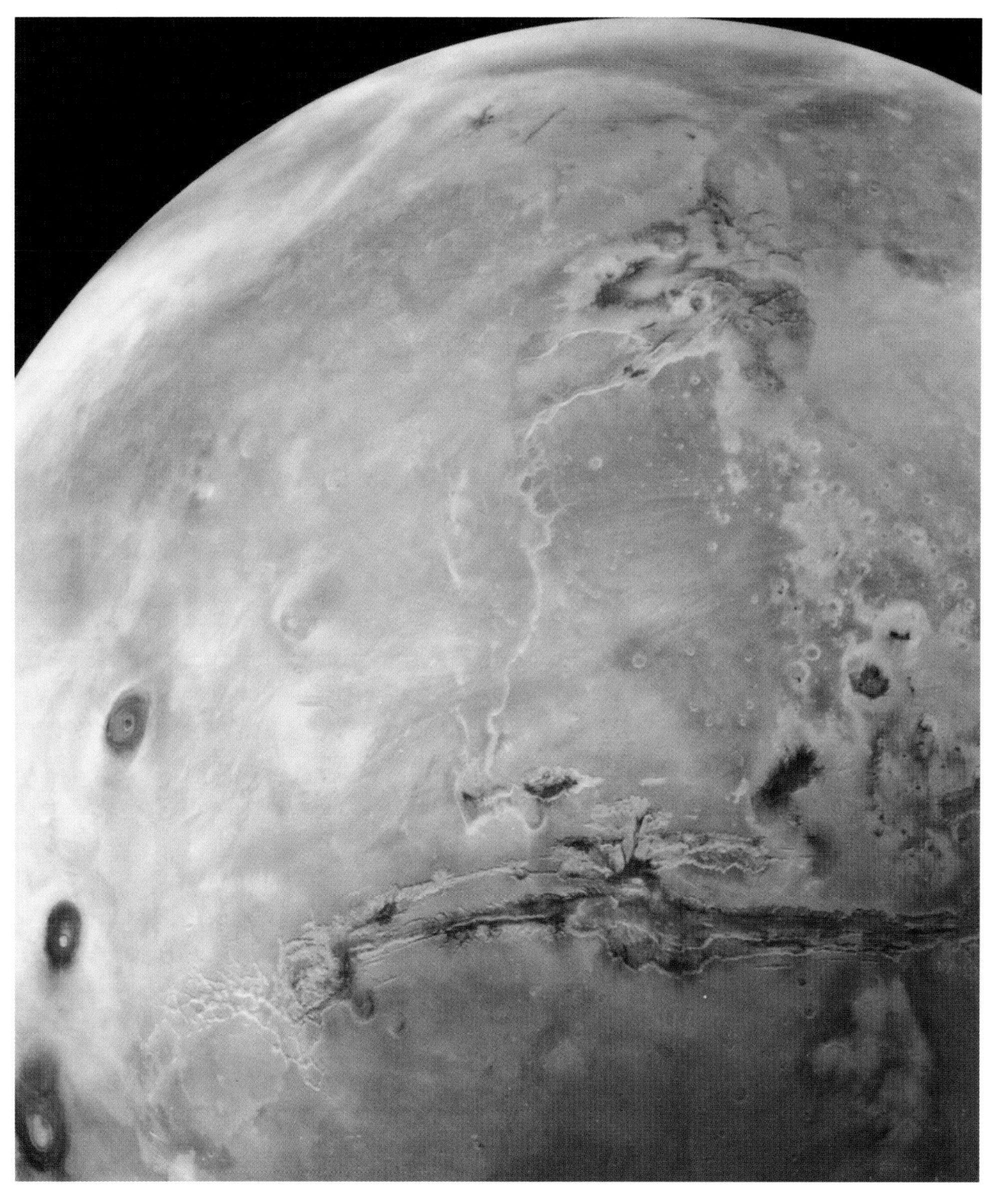

Above: A vast tract of Mars shows Valles Marineris plus three of the four largest Martian volcanoes (lower left). Facing page: The largest volcano, Olympus Mons, is three times higher than Mount Everest and bigger than France at the base. Above left: The summit caldera of Olympus Mons.

Top: On its way to a larger crater on Mars, the rover Opportunity stopped at the edge of an eight-meter-diameter crater for a day (April 23, 2004) to take this panoramic image. The crater is probably several hundred million years old. Above: Water-ice-crystal clouds at dawn on Mars, photographed in 1997 by the Pathfinder lander. Facing page: Venus global mosaic of radar scans from the Magellan mapping mission of the early 1990s.

sance Orbiter will reach Mars in 2006 carrying the best camera yet for examining the planet in high resolution. The Phoenix polar lander follows in 2008. Then a rover three times larger and more sophisticated reaches Mars in 2010.

Is there life on Mars today? The biological experiments carried out in the two Viking spacecraft that landed on the planet in 1976 yielded negative results. If life as we know it exists on Mars, it is not pervasive. Even the Earth's hardiest organisms would not survive. First, there is the cold; second, the almost total absence of liquid water; and third, a thin atmosphere that permits the Sun's ultraviolet light to stream to the surface. Unattenuated solar ultraviolet light is highly destructive to biological organisms. (One method of purifying well water utilizes a holding tank exposed to ultraviolet lamps.) Regardless, Mars remains the most Earthlike planet and the easiest to explore.

Venus: Not a Nice Place to Visit

Venus is hell; at least, that's the message we have received from robot spacecraft which have visited the planet. Not only are the surface rocks of Venus hot enough to fry eggs, but the eggs would be vaporized. A human standing unprotected on the surface of Venus would be simultaneously sizzled by the heat, asphyxiated by carbon dioxide, crushed by the density of the planet's atmospheric cloak and scorched by hydrochloric-acid vapors.

On their way to Halley's Comet in 1985, two Soviet Vega spacecraft dropped two French-designed, balloon-borne experiments into Venus's atmosphere as they flew by the planet. The experiments revealed that Venus is just as hellish in its upper atmosphere as it is at the surface. Clouds and haze far above typical jetliner cruising altitude on Earth are laden with sulfuric-acid droplets as concentrated as car-battery liquid. Two landers, released at the same time as the balloons, descended by parachute to the scorching plains of Venus, where they drilled into the surface material and collected samples for examination. The results of the experiment indicate that the rocks and soil on Venus's surface resemble those of Mars more closely than they do the Earth's.

These and other discoveries demonstrate that Venus—the only world in the solar system similar to Earth in size, mass and surface gravity—is a viciously hostile place for carbon-based life like ours. This verdict came as a huge disappointment for astronomy buffs (like me) who grew up on 1950s speculations about Venusian swamps and oceans. But such suggestions were pure guesswork on the part of frustrated astronomers who had spent decades peering through telescopes at Venus's impenetrable cloud deck. The first hint of the broiling surface conditions on our sister planet dates back to radio telescope readings in the late 1950s. Confirmation came in 1962 from instruments aboard the U.S. flyby probe Mariner 2. Because of an atmospheric greenhouse effect, temperatures on Venus from equator to poles exceed the melting point of lead.

Bathed in the dull glow of daylight filtered by a suffocatingly dense atmosphere, the Soviet spacecraft Venera 14, illustration top, gathered valuable information about the surface and atmosphere of Venus in 1982. The 460-degree-C surface temperature sizzled the well-protected spacecraft, which ceased functioning after an hour in the hellish environment. Right and facing page: Using radar to penetrate the planet's cloud cover, the Venus-orbiting Magellan spacecraft, above, revealed the true surface, as shown on these two pages in images reconstructed from the radar data. The lighter areas are the roughest surfaces.

Surface landers which followed in the years after Mariner 2 found that Venus's atmospheric blanket is 90 times as dense as the Earth's—similar in density to water at a depth of one kilometer—and is 300 Celsius degrees hotter than boiling oil.

The bad news about Venus was difficult to accept. In 1965, planetary astronomer Carl Sagan offered a brighter picture for future generations: Venus might be transformed into a more Earthlike environment through some form of global atmospheric cultivation.

"It is not inconceivable," Sagan wrote, "that an organism can be found or developed which will live and thrive somewhere on Venus and in time make it habitable for higher forms of life."

Sagan went on to propose that the clouds of Venus could be seeded with microorganisms, which "would live a wholly aerial life suspended by the turbulence that stirs the clouds." The most likely choice for such an organism is blue-green algae. The algae would transform Venus's atmosphere by extracting water and carbon dioxide, breaking them down through photosynthesis and releasing oxygen into the atmosphere. Eventually, Sagan continued, the surface would cool to a temperature that could support hardy plants and animals from Earth. He concluded: "It may even become suitable for sustaining human colonization."

But would it work? A 1982 analysis by James Oberg, then at NASA's Johnson Space Center, pointed out that the project would be a herculean effort far beyond our present or projected technology. Oberg also concluded that simple algae seeding would not work. "Assuming we want to intervene in nature," he said, "there are major roadblocks such as the spin of Venus, which is too slow for comfort."

One day on Venus is equivalent to 59 Earth days and is followed by an equally long night. If Earth rotated that slowly, life would be impossible because of the enormous heat buildup during the day and the extended nightly deep-freeze.

Capturing and flinging comets into our sister

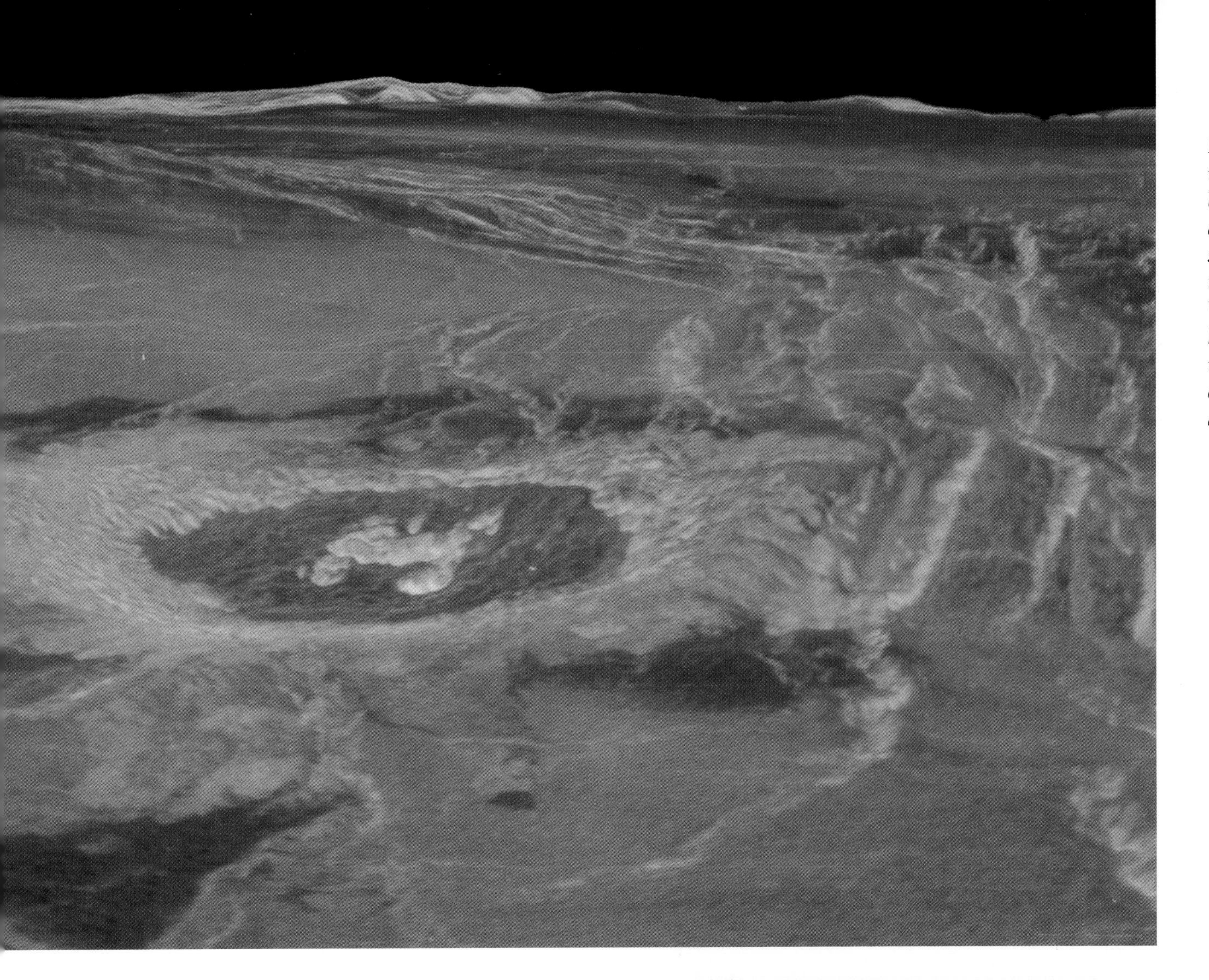

In these Magellan spacecraft images of Venus, the yellow-brown surface coloration is what an explorer would see in normal sunlight filtered through the thick atmospheric haze. The real Venus day sky is an unbroken, hazy, bright yellow curtain, but it is shown black here for visual contrast. The vertical scale has also been exaggerated for clarity.

planet has also been suggested, since they could provide water as well as accelerate Venus's rotation (if properly targeted). However, according to Oberg, the momentum involved in such collisions would not significantly alter the spin rate of Venus. Furthermore, the heat generated by the impacts would be additional excess energy that would have to be dissipated to bring the planet to Earthlike temperatures.

Even if the stifling atmosphere could be removed and the rotation problem overcome immediately, it would take centuries for the hot surface rock to cool from its searing 460 degrees C. Venus's lack of a natural magnetic field (because of the planet's sluggish rotation, which does not activate an internal dynamo like the Earth's) may have as-yet-undetermined detrimental effects on biological activities. The role that the Earth's magnetic field plays in protecting life is only partly understood. If all the obstacles could be surmounted and Venus were to be successfully transformed into an Earthlike environment, the favorable

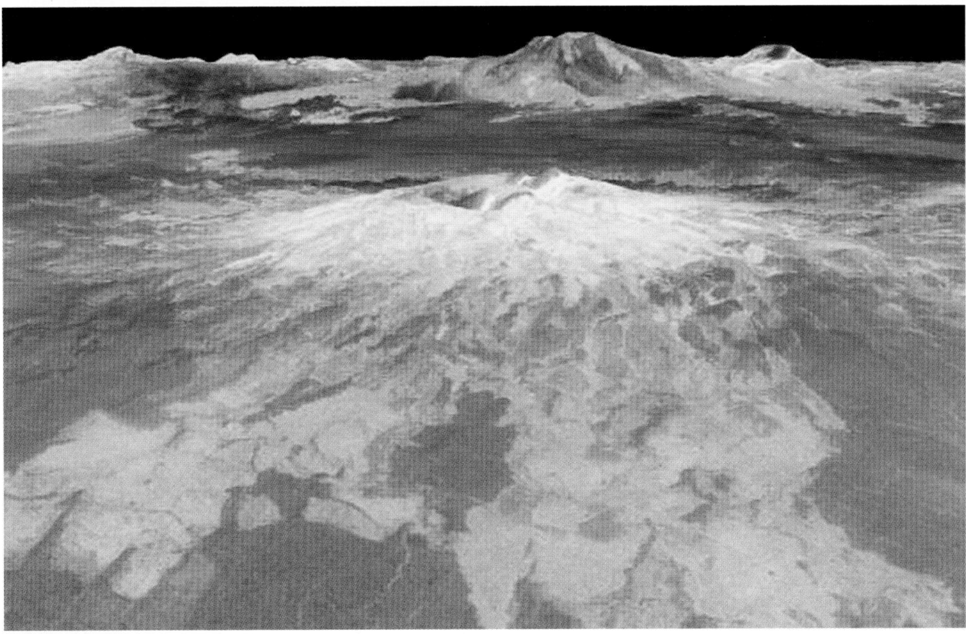

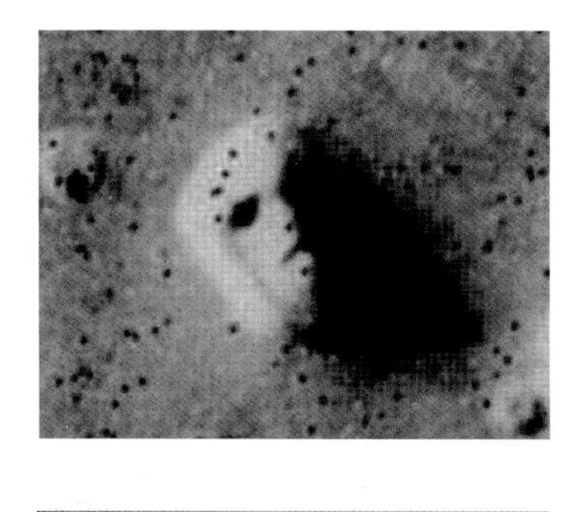

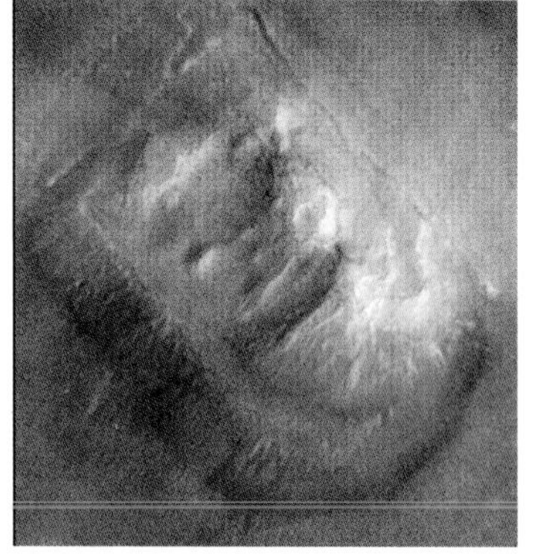

The famous two-kilometer-long "face" on Mars, top, was discovered during examination of one of the thousands of images of the desert planet taken by the Viking orbiters in 1976. This original image includes dozens of black dots caused by data "dropouts" that commonly occurred during image transmissions from the Viking orbiters. One of the dots happened to be perfectly positioned to create a "nostril," adding to the humanoid features. Claims that an alien civilization erected the monument were put to rest in 1998 when the Mars Global Surveyor spacecraft obtained the above image. With 10 times the resolution, the new image revealed that the feature is actually a craggy mound which just happens to look like a face at low resolution. Right: A region of Mars that resembles the Moon.

conditions would soon deteriorate, because the intensity of sunlight that caused the original situation would be constantly acting to reverse the process.

Why So Unlike Earth?

The swamps of Venus and the canals of Mars, described so frequently throughout the history of science fiction, have been casualties of the space age. They were wonderful dreams, but popular culture has relegated them to the attic along with Hollywood's heyday, when cheap black-and-white science fiction movies were cranked out for the drive-in market.

About the same time, a new concept dubbed the "ecosphere" made its entry into astronomical thought. The ecosphere is defined as the region around the Sun or another star where a planet similar to Earth would experience surface temperatures between the freezing and boiling points of water—in essence, a star's habitable zone, a place in front of the stellar hearth where life as we know it would be neither frozen nor sizzled.

When the ecosphere was first applied to our solar system (circa 1960), it extended from just inside Venus's orbit to beyond the orbit of Mars, with Earth wheeling through the region's midsection. With three out of nine planets in the comfort zone of our star system, the idea took root that every star similar to the Sun probably has a couple of planets tucked comfortably inside its ecosphere.

The ecosphere concept received favorable play in 1960s and 1970s astronomy books, and it led to a whole generation of both astronomers and astronomy buffs concluding that life-bearing planets ought to be commonplace in the universe. Moreover, Mars and Venus, it seemed, should be good candidates for harboring some form of life. I don't think it is any coincidence that at about the same time, the first articles on methods of detecting radio communication from extraterrestrials were published in respected scientific journals (more on this in Chapter 8). The problem—and this is nothing new—is that the theory was naively Earth-centered.

We now know what happens when a planet initially almost identical to Earth is positioned at Venus's distance from the Sun. The result is not a tropical Earth but the choking greenhouse of Venus.

Just how close could Earth be to the Sun before the runaway greenhouse effect would take over? Studies during the past few decades suggest that distance alone has very little to do with it. The mass of the planet, its volcanic history, the tilt of its rotation axis, the stabilizing effect of a large moon on the rotation axis, and so on, all play a role in planetary evolution. In the case of Mars, for example, we are dealing with a body about half the size of Earth but only one-tenth its mass. Worlds less massive than our planet have trouble holding on to an atmosphere, since their lower gravity allows more atoms and molecules of gas to escape into space. Atmospheric escape from less massive planets is further accelerated by solar radiation. But most important is evidence gathered in the past two decades which shows that a planet's atmosphere evolves over time and the Sun varies in brightness over time as well.

The global climates of Venus, Earth and Mars have all been affected by the volcanic gases each planet has emitted and by countless impacts from comets and asteroids. But the Sun's brightness evolution has had an even greater influence.

Shortly after the Sun's birth 4.6 billion years ago, it stabilized into a hydrogen-burning star known as a main-sequence star. At that time, it was shedding light at about 65 percent of its present level. Since

The first functioning spacecraft to land on the surface of Mars arrived in 1976. This shot was taken from the second lander, Viking 2. In the foreground at right is an ejected protective cover from one of the Viking instruments. The small ditch was carved by a robotic arm as it scooped up some Martian soil for examination. The on-board experiments were designed to detect evidence of life as we know it. Nothing promising was found.

In 1996, NASA scientists made a stunning announcement: the discovery of fossil bacteria inside a meteorite from Mars. The wormlike structure in this electron-microscope image of a small section of the interior of the meteorite was presented (along with other evidence) in support of this claim. The four-billion-year-old rock was apparently ejected into space when a small asteroid crashed into Mars. Eventually, the fist-sized chunk arrived on Earth as a meteorite and was discovered in pristine condition in Antarctica. The NASA team examined it for years before making its announcement. Since then, most scientists have ultimately disagreed with the NASA team's conclusions, saying that the "fossils" represent chemical, not biological, activity. The issue will probably remain controversial until Martian soil samples can be examined in laboratories on Earth.

then, the Sun has slowly but steadily increased its radiative output to the level we enjoy today.

A planet that initially experiences optimal conditions at a certain distance from its parent star might become sterile after a few billion years as the star cranks up its thermostat during its lifetime. Conversely, a planet gripped in icy cold may be thawed out as the warming star ages and evolves. But according to the fossil record, life has been abundant on Earth for at least 3.6 billion of its 4.6-billion-year history, from a time when the Sun was two-thirds its present luminosity. The Earth's average distance from the Sun could not have changed, since its orbit is extremely stable. So how did our planet avoid a global ice age?

The answer appears to be planetary and atmospheric evolution. Since more than a dozen space probes have now landed on Venus and Mars, enough detail is known of their surface conditions for scientists to piece together a picture of planetary evolution that may explain how Venus, Earth and Mars came to be so different.

First, the concept of a fixed ecosphere, or habitable zone, around a star must be discarded, both because our Sun's luminosity has increased and because a planet itself can evolve over time—especially the insulating properties of its air, which change with variations in atmosphere composition. Atmospheres evolve through volcanic activity, through sunlight decomposing various atmospheric molecules and through the escape of certain gases from the gravitational pull of the planet. In addition, there is a mixing of gases with the surface rocks and liquids. The picture is far more complex than ecosphere proponents ever imagined.

The surface of Venus is as hot as the inside of a self-cleaning oven. But in the early days of the solar system, when the Sun was dimmer, Venus's surface was probably cool enough for water to flow in rivers and to form oceans. Support for this scenario emerged in 1978 when the U.S. Pioneer Venus spacecraft sent several probes into the planet's atmosphere. Analyses of their measurements revealed that there are residual gases in the atmosphere which would be most easily explained by the existence of a global ocean of water for several hundred million years early in Venus's history. With the Sun about two-thirds its present brightness, conditions on Venus might have been similar to those on Earth today.

Whether life got a toehold on Venus at that time is unknown, but as solar radiation slowly increased to present levels, the Venus ocean was doomed. Evaporation accelerated, filling the atmosphere with water vapor. Theoretical analysis of the composition of the original atmospheres of the inner planets suggests that large amounts of carbon dioxide were present on Venus, Earth and Mars. Since those early days, the level of atmospheric carbon dioxide has fallen on Earth (today, one-tenth of 1 percent of the Earth's atmosphere is carbon dioxide) but not on Venus. Active evaporation of the primordial ocean on Venus added water vapor to the carbon dioxide in its atmosphere. Carbon dioxide alone creates an atmospheric blanket that acts like the roof and walls of a greenhouse, trapping infrared radiation. Water vapor is even more efficient at producing the greenhouse effect, heating the planet further.

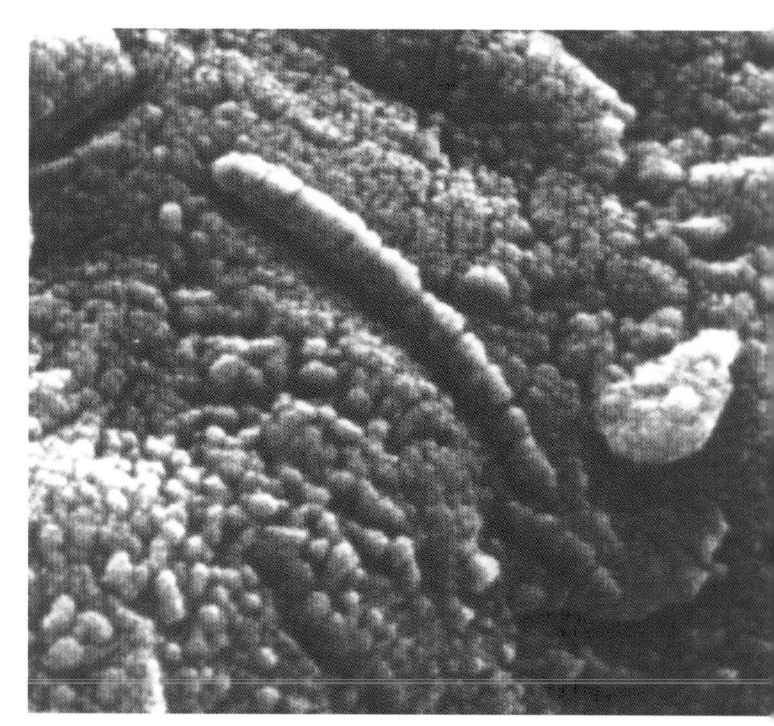

As the temperature soared on Venus, the gradually escalating solar input added fuel to the fire, heating the surface to levels that melted some of the rocks, which sent more carbon dioxide into the atmosphere and continued to drive up the temperature. Volcanoes introduced additional carbon dioxide from Venus's interior along with other gases, and the greenhouse cycle spiraled until the planet reached its present temperature (460 degrees C), which exceeds the melting point of lead. Today, Venus's atmosphere contains about 300,000 times as much carbon dioxide as does the Earth's.

Research combining the results of the Venus-orbiting Magellan mission of the 1990s with findings from the Pioneer Venus orbiter a decade earlier has reinforced these conclusions. Magellan's radar maps show a world that was literally repaved by a kilometer-thick layer of lava around 800 million years ago. Large quantities of water vapor and sulfur dioxide, both greenhouse gases, must have been spewed into Venus's atmosphere by this era of colossal volcanic activity. Over the following 800 million years, there may have been periods of tens of millions of years when the planet's surface temperature was 100 Celsius degrees higher than today's blistering conditions. Furthermore, volcanoes are almost certainly still active on Venus today.

The haze and cloud decks in Venus's upper atmosphere reflect sunlight more effectively than do the Earth's clouds, so the amount of sunlight reaching the surface of Venus is less than Earth receives on a cloudy day. Hypothetically, if the clouds could be retained but the greenhouse effect eliminated, Venus would be cooler than Earth. This is wishful thinking, though. Venus is trapped. With the Sun continuing to warm it and no liquid water to cycle Venus's carbon dioxide back into carbonate rocks, the planet is a smothered inferno.

On Earth, however, the initial elevated level of carbon dioxide was exactly what was needed to keep our planet from becoming locked in a worldwide ice age. The key to the Earth's survival was that it never became warm enough for the surface water to start boiling. The oceans contain more than 50 times as much carbon as the atmosphere, mostly dissolved in the form of bicarbonates. If the oceans ever were to boil, the oxygen in the water would combine with the carbon to form carbon dioxide in huge quantities—as must have happened on Venus—and the hydrogen would escape into space.

Apart from the essential direct role water plays in life processes, oceans keep carbon dioxide from building up by absorbing it from the atmosphere and then cleansing it from the water through the production of carbonate rocks and seafloor sediments. Rain, which contains dissolved carbon dioxide from the atmosphere, aids the cycle when it enters the oceans. Once there, carbon is taken up by marine organisms. When the organisms die, their shells sink to the bottom and form limestone. The reverse process occurs when a volcano spews carbon dioxide formerly trapped in the rocks back into the atmosphere or into the oceans from the volcanic midocean ridges.

This can be a self-regulating process. As the Sun gets hotter, it evaporates more water from the oceans, which causes more rain, thus reducing the atmospheric carbon dioxide content. When green plants appear, they remove more carbon dioxide from the atmosphere and oceans and return oxygen, which is nowhere near as effective an insulating atmospheric gas as is carbon dioxide. Warm and cool cycles in the Earth's climatic history may be directly related to changes in carbon dioxide levels. These, in turn, can be traced to plate tectonics (continental drift). When plate motion is most dynamic, more carbon dioxide

Venus is the nearest planet and the brightest starlike object visible from Earth in the nighttime sky. In this photograph, taken in February 1999, Venus appears along nearly the same line of sight as Jupiter, the second brightest planet.

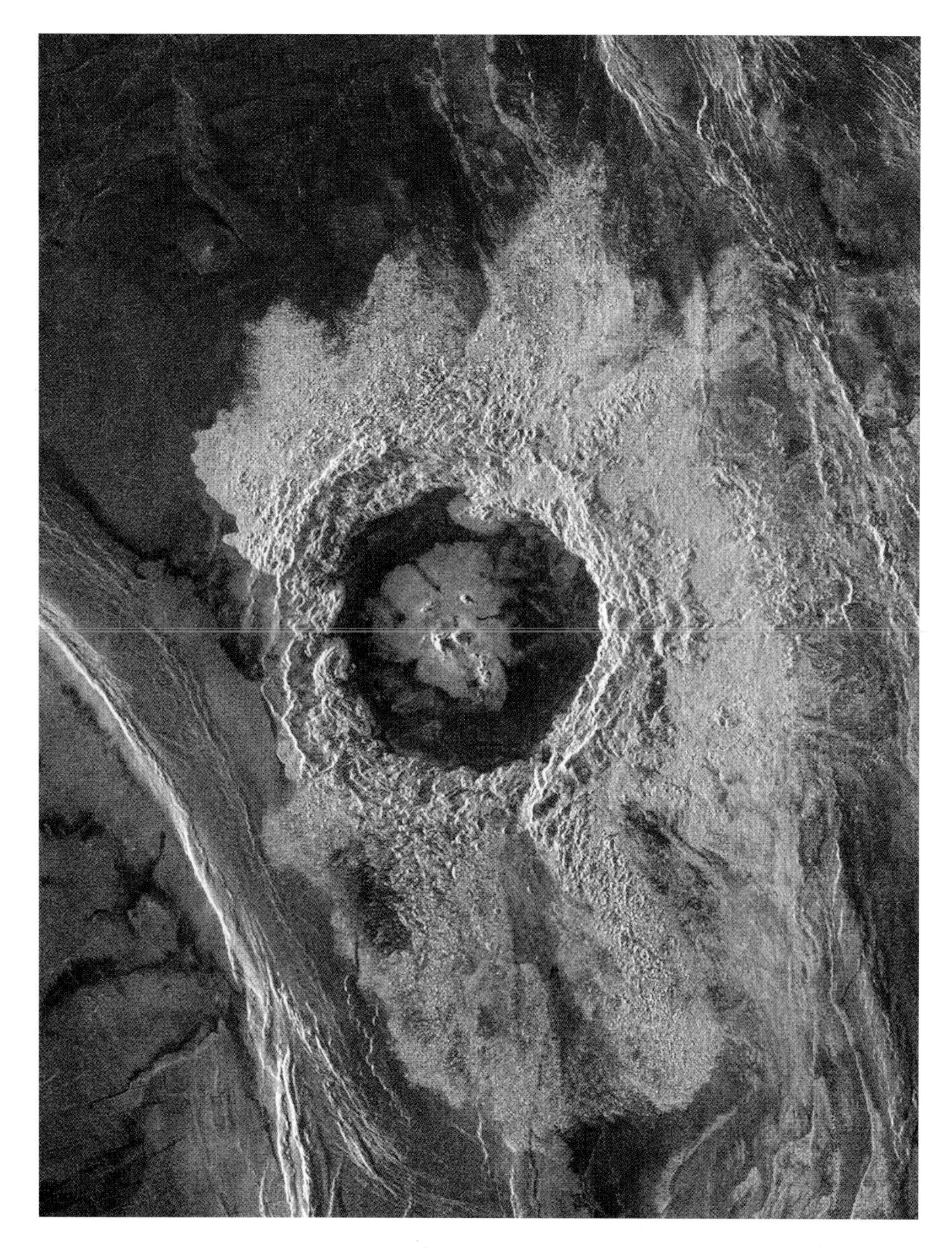

One of the most prominent craters on Venus, at 185 kilometers in diameter, crater Dickinson was named in honor of American poet Emily Dickinson. The surface features of Venus are all named for historically prominent women from the arts, the sciences and mythology.

is introduced into the atmosphere; the temperature then rises, producing climate regimes such as the lengthy warm spell about 100 million years ago that marked the height of the Age of Dinosaurs, when most of the planet experienced tropical conditions.

In about four billion years, the Sun will release 50 percent more energy than it now does, the result of a shifting of its internal heat-producing mechanisms into "shells" that will surround the core. Mars will then receive enough solar heat to begin to thaw the global deep-freeze. It is a long time to wait, but when it happens, the permafrost will melt and liquid water may flow once again on the surface of Mars. The more Earthlike environment could last for millions of years, just as it probably did on Venus four billion years ago. It might allow some form of life to arise on Mars, but that is sheer conjecture. What seems more definite is that if Earth and Mars were to change places today, the Earth's oceans would freeze and the planet would be locked in a global ice age that would not eliminate all life but would certainly reduce its range of environments.

Conversely, if Mars were cruising in the Earth's orbit, the permafrost in the equatorial region would melt. Whether the added atmospheric water vapor would be adequate to offer life-support environments is unknown. But this kind of speculation is littered with caveats, because we simply do not understand the conditions necessary for life's genesis or the branching paths that life will take once it gets a foothold on a world. We are stuck with only one example: Earth.

Mercury: Sun-Scorched World

An astronaut trained for exploration of the Moon would be well prepared for the rigors of the innermost planet of the solar system. Mercury is an airless, cratered world half again as large as the Moon but still only one-third the diameter of Earth. Astronauts on Mercury would feel about twice the Moon's gravitational pull but just one-third of that experienced on Earth. With the Sun shining six times as intensely as it does at noon on Earth, efficient spacesuit air-conditioning would be a prerequisite.

Of all the worlds in the solar system explored so far, Mercury proved to be the least surprising when the Mariner 10 spacecraft provided our first (and only) close-up views in 1974. Mercury's face, like the Moon's, still shows the scars of saturation bombardment by asteroids and comets during the solar system's youth. For the last three billion years at least, it has remained essentially unaltered.

The planet's proximity to the Sun and its small size made it an exceedingly difficult object to study when the only available tools were Earth-based telescopes, and even the most powerful of those never revealed more than a few pale smudges on Mercury.

And although the Hubble Space Telescope is the premier astronomical tool available, it can't be used to look at Mercury because the orbiting telescope

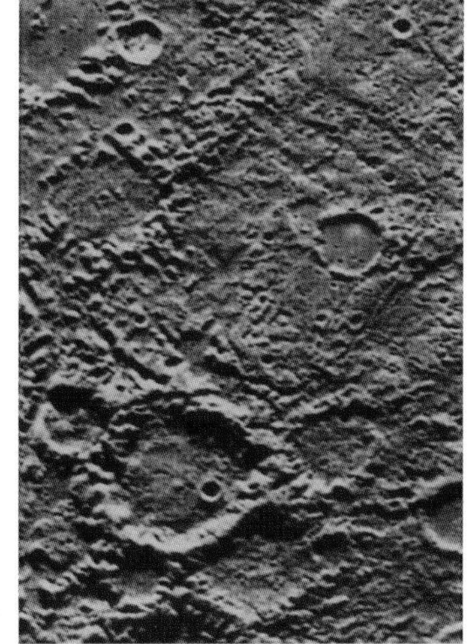

Mercury's sun-baked surface, similar in many respects to the surface of the Moon, has changed little over the past three billion years. These two portraits of Mercury's cratered surface were captured by the cameras on board the U.S. Mariner 10 space probe that swung by the planet in 1974. The next mission to the innermost planet is the Messenger spacecraft, scheduled to reach Mercury in 2008 to begin mapping the entire planet over a three-year period.

would have to be pointed too close to the Sun. The Mariner 10 mission was a great success, but the spacecraft trajectory limited the views to only one side of the planet.

After many proposals to return to Mercury, it looks as though it is finally going to happen with launch of the Messenger spacecraft in 2004. The plan is to take a convoluted but highly energy-efficient trajectory involving Earth and Venus. After launch, Messenger will return a year later for a flyby of Earth at a different angle, which will aim the spacecraft toward Venus. After a flyby of Venus, Messenger will return to Venus for a second flyby at a different angle and then finally head toward Mercury in 2008. At this point, Messenger will begin a three-year mapping program from a steep elliptical orbit.

Until the early 1960s, textbooks stated as fact that one side of Mercury constantly faces the Sun, leaving a perpetual twilight zone around the sunrise-sunset transition, where some optimists suggested life might be found. Then, in 1965, radar signals bounced off Mercury revealed what more than a century of telescopic observation from Earth had not: One side of Mercury does *not* always face the Sun. In a bizarre slow-motion waltz, the innermost planet rotates on its axis three times in exactly the time it orbits the Sun twice. The combination of these two motions results in a true day (sunrise to sunrise) that is 176 Earth days long, twice the length of Mercury's year!

Noon temperatures at Mercury's equator reach about 400 degrees C, but because there is no insulating atmosphere, a spacesuit air conditioner could radiate into the vacuum of space the appropriate amount of absorbed heat to keep an astronaut cool.

Mercury is believed to be the most mineral-rich major body in the solar system, although I doubt whether this will be used as a reason to send astronauts to the planet. If a pure mining expedition is ever contemplated, billions of tons of the same minerals, in the form of passing asteroids, fly much closer to us. An asteroid's low surface gravity makes extracting and transporting material much easier. Instead, as the inevitable bubble of exploration expands in the centuries to come, Mercury will take its turn among hundreds of other bodies in the solar system that will become targets of ever more detailed study.

REALM OF THE GIANTS

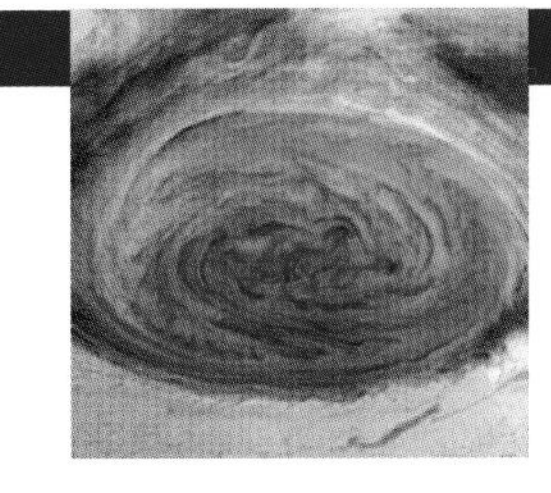

CHAPTER THREE

Hurricane-force winds whip up storms that could swallow the continent of Asia. Lightning bolts capable of vaporizing a small city rip open the sky. A swirling vortex of clouds larger than Earth has been raging for centuries and shows no signs of abating. Above it all, lethal radiation permeates the surrounding environment. This is Jupiter, a bizarre colossus that dwarfs every standard of comparison familiar to Earthlings.

Jupiter is the largest and nearest of four giant planets, all different in almost every respect from Earth and its companions in the inner solar system. The contrast between the giant planets and those close to the Sun can be traced back to the solar nebula.

For a brief but crucial period, perhaps as little as 500,000 years, the dazzling primal Sun blazed with a brightness which was at least 20 times that of the present-day Sun. Stoked by the heat of contraction piggybacked on youthful thermonuclear fires, the brilliant newborn star blasted the lighter gases from the primordial nebula that cocooned the inner solar system, but only marginally affected conditions from the distance of Jupiter outward. The composition of Jupiter and Saturn—four-fifths hydrogen, one-fifth helium—is roughly the same as that of nebulas which float in the spiral arms of the Milky Way, where stars are being born today. This indicates that these worlds coalesced from essentially pure solar-nebula construction material, not the heavy, hard-to-melt residuals which make up Earth and its neighbors. The Sun's planetary family was thus, from the beginning, divided into two camps: small rocky inner worlds and lumbering gas-fat giants in the outer regions.

Jupiter is the best known of the gas-giant planets and the most alien. Its visible surface is an endless maelstrom of clouds so vast that if Earth could be peeled like an orange, the skin would not quite cover the Great Red Spot, the planet's largest storm center. In size, this planetary behemoth is to Earth what a basketball is to a golf ball. More than twice as massive as all the other planets and their satellites combined, Jupiter has no rivals in the solar system.

The big planet's quilt of ivory- and salmon-colored clouds is a facade, an ever-changing mask of many hues that blankets the real planet underneath, which is essentially a cosmically sized drop of liquid hydrogen and helium 318 times the Earth's mass. Encased at the very center is a small rock-and-metal core approximately the size of Earth.

Seen from Earth, Jupiter is the fourth brightest object in the sky after the Sun, the Moon and Venus. Ponderously navigating its huge 12-year orbit about the Sun at five times the Earth's distance, Jupiter drags with it a miniature solar system of at least 61 moons, two of which are the size of the planet Mercury. The giant planet completes one rotation on its axis every 10 hours, making its day the shortest of the nine planets. A point on Jupiter's equator races along at 35,000 kilometers per hour compared with 1,600 kilometers per hour for a similar point on Earth. The tremendous rotational speed combined with the fluid character of the planet make Jupiter bulge at its equator, giving it a polar diameter about 7 percent smaller than its equatorial diameter.

Human exploration of Jupiter seems impossible, at least for the foreseeable future. In addition to the big planet's immense gravity, Jupiter's powerful magnetic field wraps lethal radiation belts into an invisi-

How vast those Orbs must be, and how inconsiderable this Earth.

Christiaan Huygens
1689

The most detailed portrait of Jupiter ever obtained was captured by the Cassini interplanetary spacecraft as it passed the giant planet in December 2000 on its way to Saturn. This picture is a composite of 27 high-resolution frames. Two years of painstaking alignment and processing of individual jigsawlike images were required to create this single image. Above: Galileo spacecraft view of Jupiter's Great Red Spot.

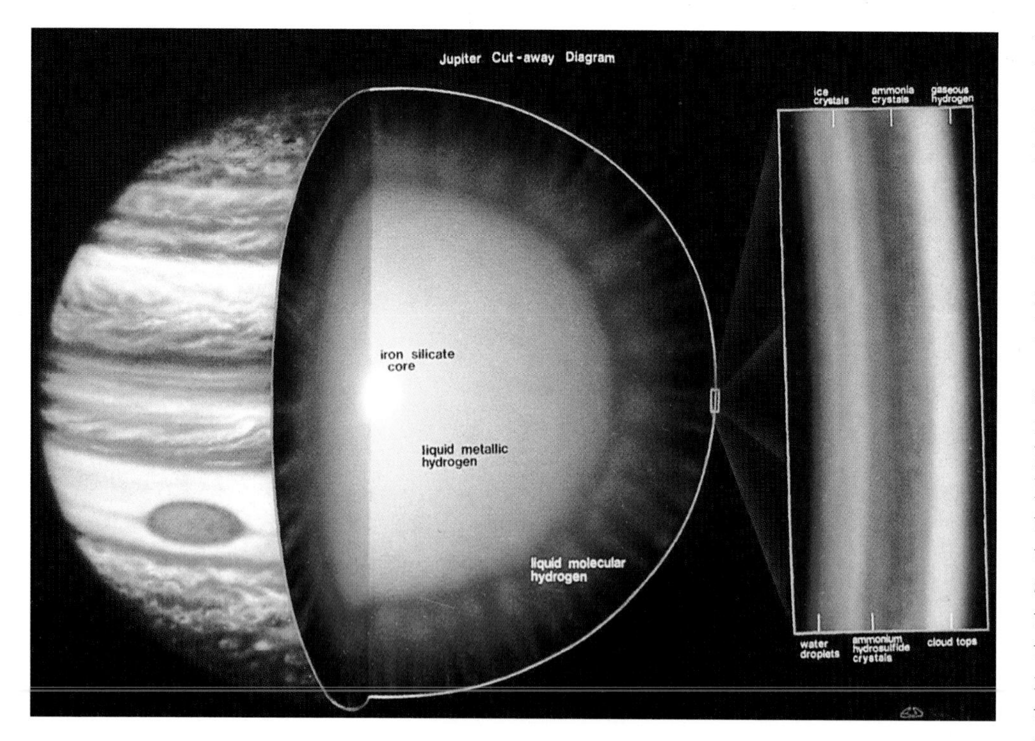

Above: Jupiter's solid rocky core, perhaps 10 times the mass of Earth, marks the center of the giant planet. The overlying layers of hot, compressed hydrogen would crush any conceivable robot probe. The "surface" clouds are accessible to exploration —but only with difficulty. Right: This Galileo spacecraft image happened to capture a Texas-sized lightning storm raging in Jupiter's clouds. Facing page: Its nuclear furnace newly ignited, our youthful Sun pumped an intense surge of radiation past the rocky inner planets, including Earth, sweeping the zone clear of gas and dust and leaving these worlds airless. Protected by distance, the outer planets, like Jupiter, were largely unaffected by this phase, which lasted for about 10 million years, and went on to gather and retain massive gaseous atmospheres.

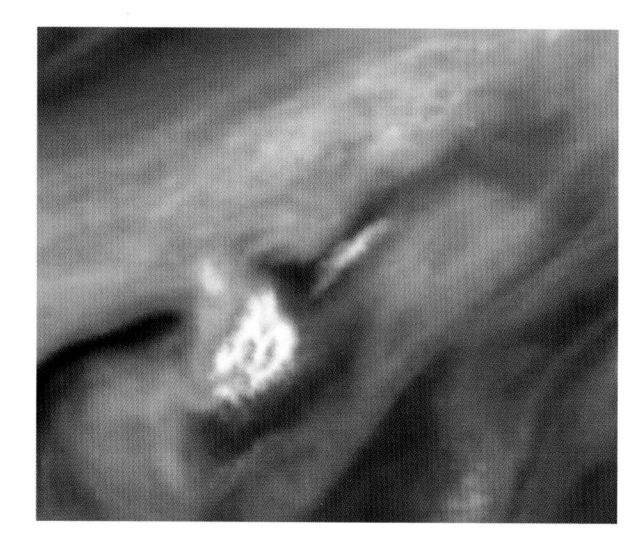

ble shield around the planet. Someday, though, a device combining the features of an interplanetary spacecraft and a submarine may take humans on a journey into the Jovian clouds. Approaching Jupiter's ocean of clouds, the crew of such a vehicle would descend first through rarefied haze layers of ethane and acetylene before moving into the clear hydrogen-and-helium atmosphere with its tiers of streaming white clouds, similar in appearance to cirrus clouds but consisting largely of ammonia ice crystals. These clouds would become thicker and more billowy as the descent continued, until nothing would be visible but white sheets swirled into vortexes and occasionally skewered by cloud towers from below. At this level, an instrumented exploration probe suspended by a balloon inflated with heated hydrogen could ride the winds indefinitely. The temperature here is minus 120 degrees C, and the atmospheric pressure is about 70 percent of that on the Earth's surface. Twelve kilometers above, at the 30 percent atmospheric-pressure level, the temperature is minus 145 degrees C.

Twenty kilometers below the tops of the white ammonia cloud decks, a transition zone leads to multicolored ammonium-hydrosulfide-crystal clouds: beige, peach, khaki, salmon and brown. They form horizontal belts around the planet that alternate with the higher zones of ammonia clouds and constitute the two main features of Jupiter's visible face. The relatively permanent colored belts and white zones are the planet's weather system, similar in some ways to the Earth's. But in the Earth's case, the weather machine is powered by the Sun, whereas on Jupiter, the system is driven more by internal planetary heat than by solar radiation.

On Earth, huge masses of warm, light gas rise to high altitudes, cool, become heavier, then roll down the sides of new rising columns of gas. The force of the Earth's rotation creates an overriding west-to-east flow. Instabilities convert the motion into the enormous spiral storm centers (cyclones and anticyclones) seen in weather-satellite images. The Earth's weather system is further complicated when air encounters continents and mountain ranges, which disturbs the atmospheric flow.

On Jupiter, the planet's more rapid rotation virtually eliminates north-south flow, whipping the clouds into distinct alternating high- and low-pressure bands parallel to the equator. Spiral storms are quickly twirled into continent-sized oval disturbances that turn like ball bearings between the main atmospheric bands. The biggest of these storms, the Great Red Spot, is the highest feature on Jupiter and is composed of material dredged up by an especially powerful cyclonic system. How it has sustained itself for so long is a mystery (drawings made in 1664 show it), but without landmasses to bump into, it is difficult for such a large storm to dissipate once it gets going. Because it intrudes well into the two counterflowing currents in the atmospheric wind zones to its north and south, the Red Spot may tap their energy of motion more effectively than a smaller vortex could.

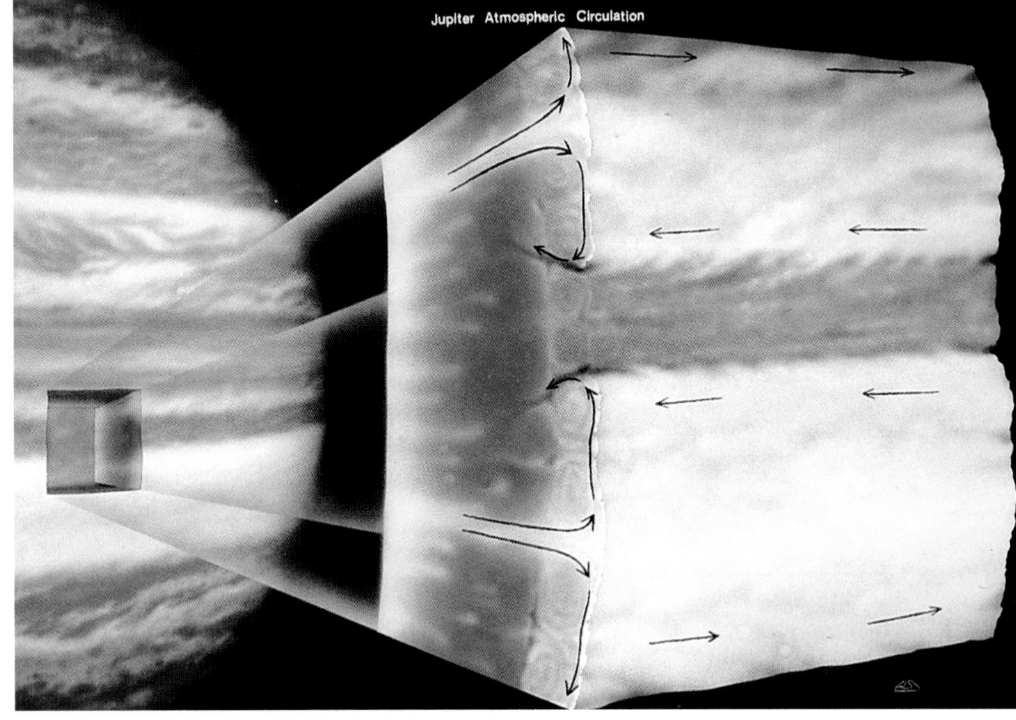

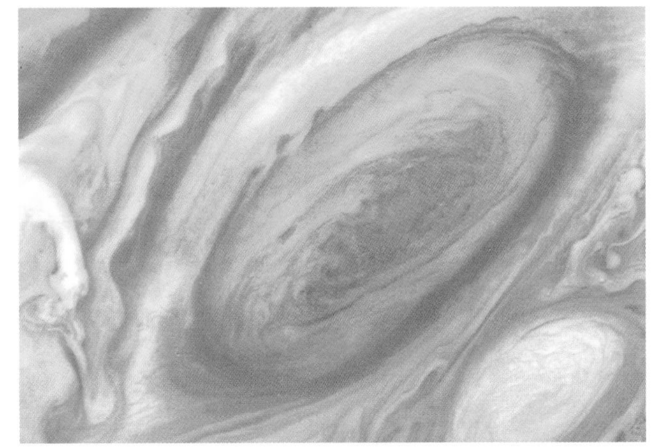

Nowhere in the solar system is the cloudscape as active or as colorful as on Jupiter, where booming winds and powerful vertical drafts whip ammonia and ammonium-hydrosulfide clouds into an ocean of unearthly beauty. Rendering at far left is a view from near the clouds' "surface." The clear hydrogen air allows sightlines hundreds of kilometers long on this largest of the planets. Above it all, Jupiter's moons Io and Europa are dual crescents in the sapphire sky. The major atmospheric feature, known as the Great Red Spot, center, is a whirlpool-like storm larger than Earth.

Phantom-thin, Jupiter's rings, top, are so elusive that they were not discovered until the Voyager 1 spacecraft reached Jupiter in 1979. This Galileo spacecraft image was significantly enhanced for publication. Above: Saturn's rings were illuminated from above as Voyager 2 swung below them in 1981.

Not far below the ammonium-hydrosulfide clouds, the temperature is high enough that ammonia is no longer frozen, and water ice crystals drift around in the weather circulation along with ammonia droplets. A few kilometers farther down, the temperature rises above the freezing point of water. Frost and ice accumulation on the exterior of an exploration vehicle would melt at this level, where water is abundant in the form of alkaline droplets containing ammonia in solution. The ammonium-hydroxide mist here was once thought to be a lethal environment for all forms of life, but bacteria have been found on Earth in hot springs far more alkaline than was considered possible only a few years ago. Other bacteria can withstand temperatures higher than the boiling point of water, and some do not require oxygen. There is no doubt that these simple forms of life thrive on Earth under conditions which are normal for Jupiter. The question is, Does life exist in Jupiter's atmosphere? Recent evidence suggests that it does not.

Although not intended to answer the life question, the four Pioneer and Voyager probes that sailed by the giant planet during the 1970s and the Galileo probe that orbited Jupiter in the 1990s provided an unprecedented examination of the eddies and currents in the Jovian atmosphere. Strong vertical motion seems to be prevalent, even violent, in some regions. The constant stirring could, within hours, carry atmosphere-borne life through a temperature range of hundreds of degrees. While this, in itself, does not eliminate the possibility of life in Jupiter's atmosphere, it is a discouraging sign.

The ammonium-hydrosulfide clouds form a ceiling above the potential life zone, where the atmosphere is misty but essentially clear. Below looms an abyss leading to total blackness. From this level down, our exploration takes on the characteristics of a deep-sea dive, as the atmospheric pressure escalates to more than 100 times the surface atmospheric pressure on Earth and sunlight is dimmed to near-nightfall levels. The pressure increase continues until the equivalent of seawater pressure in the deepest ocean trenches on Earth is crushing around our space capsule. The descent is eventually stopped 1,000 kilometers down by rising currents of liquid hydrogen denser than lead. At 2,000 degrees C, it is hotter than the inside of a blast furnace. The temperature at a depth of 3,000 kilometers is 5,000 degrees C, and the pressure is 90,000 times the atmospheric pressure on the Earth's surface.

One-third of the way to Jupiter's core, the temperature reaches 10,000 degrees C and the pressure is three million Earth atmospheres. At this level, liquid hydrogen turns into liquid metallic hydrogen, which has the properties of metal and is responsible for Jupiter's enormously powerful magnetic field. Deep inside Jupiter, near the core, incredible pressures are concentrated in a molten region where the temperature is 30,000 degrees C, five times hotter than the surface of the Sun. Yet this does not even begin to approach the several million degrees required to ignite the thermonuclear "fires" that power stars.

Jupiter has been called the star that failed. While it is true that the giant planet is composed of fundamentally the same ingredients as a star, it was never close to being one. To become a true star and fire up its nuclear furnace permanently, Jupiter would have to be 80 times more massive. Nevertheless, the planet's core is hot enough to generate seething convective currents that, in about a century, transport heat from the interior to the upper atmospheric levels, where the rise and fall of cloud currents finally disperse it into space. The energy released is substantial: Jupiter radiates $2\frac{1}{2}$ times as much heat as it receives from the Sun. The best explanation for the existence of the high internal temperature is that it is maintained by a slow contraction of the planet. With such a huge mass, a small contraction goes a long way. A shrinkage of just a few mil-

Girded by its magnificent rings, Saturn is one of nature's most breathtaking spectacles. This 2004 Hubble Space Telescope image is the finest portrait of the planet taken by a telescope at the Earth's distance.

limeters per year can account for the planet's tremendous internal furnace.

Above Jupiter's highest cloud deck, the environment is just as unearthly as it is below. A deadly invisible shield—Jupiter's magnetic field—surrounds the planet. The total energy of the Jovian magnetic field is 400 million times the Earth's comparatively puny field. Jupiter's radiation belts, embedded in the field, are up to 100 times stronger than the dose lethal to humans and are near the tolerance limit for spacecraft systems. No other planet comes close in this respect, making Jupiter and its inner moons forbidden territory for human exploration for decades to come.

Among the more surprising of the many Voyager discoveries at Jupiter was a ring—a thin disk-shaped swarm of tiny particles—surrounding the planet. Made up mostly of dustlike bits of matter that become visible only when illuminated from behind by sunlight, the ring was captured in a photograph as Voyager passed Jupiter's nightside. The structure is too faint to be seen from Earth, although once astronomers knew what to look for, they were able to detect it using electronically enhanced telescope imaging systems. The ring extends all the way down to the planet. Over time, pressure from sunlight and interaction with Jupiter's magnetic field cause the ring particles to spiral slowly inward and ultimately settle in Jupiter's upper atmosphere. The source to replace the lost particles is both the inner small moons of Jupiter, whose surfaces are gradually being worn down through constant contact with orbiting debris in their vicinity, and Jupiter's moon Io, which has its own amazing way of expelling material into space. (The moons of the giant planets are described in Chapter 4.)

Saturn: Lord of the Rings

Saturn, the second largest planet in our solar system, is almost identical to Jupiter in composition and structure but is somewhat cooler and less active in its interior. Because Saturn is more remote from the warming influence of the Sun, a high-altitude haze layer of cirrus-type clouds of ammonia ice crystals—much thicker than Jupiter's—veils Saturn's face, producing a much blander appearance. The subdued interior heat generates less turbulent surface eddies, so the overall effect is a quieter atmospheric circulation. Although less impressive than Jupiter in size and surface features, Saturn is endowed with a hauntingly beautiful ornamentation—its rings—one of the universe's most dazzling creations.

Saturn's rings are made up of swarms of icy particles ranging in size from tiny crystals, like those in an ice fog, to huge flying icebergs as big as small

mountains. Each of the ring particles has its own individual orbit about Saturn, although a gentle jostling occurs as the particles are affected by the gravitational pull of Saturn's major moons and of each other.

The rings are truly enormous in extent. From one edge to the other, they span a distance equivalent to two-thirds of the gulf between Earth and the Moon. Yet the particles that make up the rings seldom stray more than a few hundred meters from a perfectly flat disk, making the structure about as thick as the height of a 30-story building. A scale model of the rings made of paper the thickness of the page you are now reading would be larger than a football field.

The reason the rings are not spread in a random haze around Saturn is related to the distribution of mass within Saturn itself. Saturn is the least dense of the gaseous giant planets, and its rapid 10.7-hour rotation period — Saturn's day — has significantly bulged the planet at the equator and compressed it at the poles. Thus a body orbiting around Saturn's equatorial zone "feels" a greater gravitational pull than when it passes over the polar regions, because there is more material below it. The path of greatest orbital stability is a nearly circular one above the most massive sector of the planet, precisely at its equator. Within the rings, gentle collisions gradually grind down the larger particles. Meanwhile, the smallest particles tend to stick to one another and build into larger clumps through accretion. These actions have established an equilibrium between the destruction of large particles and the accretion of small ones into bigger ones. Observations from Voyagers 1 and 2 show that the very largest particles are near the rings' outer edge, the zone of the least disruptive forces.

A Ring Walk

Although the question of the origin of Saturn's rings remains unanswered, the leading theory is that either two of Saturn's moons collided or an outside force such as a giant comet smashed one of the moons apart and the pieces then evolved into the rings we see today. Recent though still controversial evidence suggests that the rings could be less than one billion years old—perhaps much less. But conventional wisdom still dates the elegant structure closer to the origin of the solar system, more than four billion years ago, a time when comet collisions were far more common than in the past billion years.

An exploration of the rings by a spacesuited astronaut outfitted with a propulsion backpack for maneuvering would be one of the most exquisitely beautiful excursions a human could take in exploring other worlds. A spaceship orbiting Saturn could bring the astronaut to within a few kilometers of the ring structure. So long as the spacecraft remained in an approximately circular path around Saturn's equator, its speed would be identical to the speed of the ring particles. A collision with any ring material under these conditions would be just a gentle nudge.

Imagine. Stepping untethered outside the spacecraft, the ring explorer uses brief bursts of backpack propulsion to make a slow glide toward the glittering golden plain of ring material extending seemingly to infinity. Descending into the ring structure, our cosmic traveler touches down on a large ring particle and, with a push of the foot, is propelled to the next major piece. In between this slow-motion ballet toe-step, the explorer moves through a blizzard of particles gently bouncing off the front of the spacesuit and faceplate. Simply floating in the rings' silvery gravel and being carried around the planet with it would be an intoxicating experience: The partly hidden Sun glinting off the multitude of particles, the gentle gravitational symphony of collective motion that carries them around the planet, the feeling of being surrounded yet freely floating in space—all disguise the fact that everything is whirling around the planet at a velocity of tens of thousands of kilometers per hour.

For every house-sized ring boulder, there are a million the size of a baseball and trillions the size of a grain of sand. In denser sections of the rings, the baseball-sized particles would be separated by a meter or so, while the house-sized ones would be relatively rare, sometimes kilometers apart. But if the entire ring structure were melted and refrozen as a solid body, the disk would be less than a meter thick.

Not only are Saturn's rings astonishingly bright, but they are also conveniently tipped by the planet's 27-degree axis tilt, allowing us a splendid viewing angle. Since the rings ride exactly above Saturn's equator, we would never properly see them from Earth if Saturn had an axial inclination of, say, three degrees, like Jupiter. Only spacecraft photographs would reveal their full glory. As it is, though, we are treated to a majestic cycle of visibility during Saturn's 29-year orbit of the Sun. The current cycle began with the rings seen edge-on in 1996. They were fully open in 2002, with the south side presented to our view, and will be edge-on again in 2009. The north side of the rings is at maximum tilt toward us in 2016, then they close to edge-on again in 2025.

Above: This enhanced false-color view of Saturn's rings emphasizes the divisions and grooves throughout the delicate structure. The Voyager 2 picture on the facing page ranks as one of the great astronomical images of the 20th century. Three of Saturn's family of moons march in the blackness in front of Saturn, while a fourth is visible closer in, appearing as a white speck near the edge of the planet at the bottom of the disk. The shadow of that moon is the smaller of the two black dots on Saturn's clouds. The larger shadow is cast by the left moon of the lower trio. The full extent of the rings from edge to edge would span two-thirds of the distance between Earth and the Moon.

Images of Jupiter and Saturn that rival those taken by research telescopes are now being gathered by amateur astronomers using inexpensive digital cameras attached to backyard telescopes.

Of course, Saturn itself is not wobbling around to create these changes. Saturn's rotation axis is essentially fixed in space over the centuries, like the Earth's. What we are seeing is Saturn's seasons—first spring and summer in the southern hemisphere, then spring and summer in the northern hemisphere.

The highly reflective particles of water ice that make up the rings are bright enough to outshine the ammonia clouds that top Saturn's atmosphere. The ring systems of the other giant planets—Jupiter, Uranus and Neptune—are drab appendages compared with the sparkling beauty of Saturn's adornment. The rings of Jupiter are dull and diffuse; those of Uranus and Neptune are downright dark, among the least reflective objects known in the solar system. If any of these ring systems surrounded Saturn, the ornament would be either barely visible or totally invisible to a visual telescopic observer on Earth. All these factors together, plus the fact that Saturn's rings are simply more massive than those of any other planet, make them a magnificent spectacle.

Although spacecraft have revealed hundreds of identifiable rings in the overall ring structure surrounding Saturn, only three components can be distinguished visually through a telescope on Earth: rings A, B and C. Ring A, the outer band, and ring B, the widest and brightest section, are both easily visible in any telescope. They are separated by Cassini's division, a broad gap about as wide as North America. Cassini's division looks as black as the sky around Saturn but is actually a region of less densely packed ring particles rather than a true blank space. The division is caused by gravitational perturbations by Saturn's moon Mimas. Ring particles orbiting in the gap have what is called a resonance with Mimas and are, over time, gravitationally nudged into new orbits, thus largely clearing out the section. The other gaps that produce the many discrete rings seen by the Voyager spacecraft are generated through more complex interactions with other moons and with large particles within the rings.

Ring C, the innermost ring, is so dim that it takes a large telescope to reveal it beside the brilliance of Saturn itself. Also known as the crepe ring, ring C is a phantomlike structure extending about halfway to the planet from the inner edge of ring B.

There is much yet to be learned about Saturn and its family of moons and rings, which is why the Cassini spacecraft was dispatched in October 1998 to reach Saturn in July 2004 and swing into orbit for a four-year exploration. Eight months after its arrival, Cassini will drop an instrumented probe named Huygens into the atmosphere of Titan, Saturn's largest moon and the only known satellite in the solar system with a substantial atmosphere (more on Titan in Chapter 4). During its four-year mission, Cassini will take tens of thousands of close-up images of Saturn and its rings and moons.

Uranus: Aquamarine Giant

Although it dwarfs Earth, Uranus is so remote that it appears as merely a pale greenish ball even in the largest telescopes on Earth, its five major moons tiny dots scattered around it. Not even the planet's rotation period—its day—was known with any accuracy prior to the first spacecraft encounter by Voyager 2 in 1986. Voyager's cameras detected several inconspicuous clouds in the planet's cloak of aquamarine "smog." Monitoring the clouds for a few rotations revealed that Uranus turns in 16.7 hours. Nothing like the colorful cloud bands of Jupiter and Saturn were seen by Voyager. The lack of cloud activity can be partly attributed to Uranus's weak internal heat engine, while a haze of methane ice crystals, a sort of ice fog, further acts to block what lies beneath. Some rare towering clouds, elevated by occasional energetic eddies from below, have been recorded by the Hubble Space Telescope—but overall, Uranus seems to have the most featureless visible surface of any planet.

Uranus is mammoth by earthly standards: 63 times larger than Earth in volume, 14 times its mass and 4 times its diameter. A rocky structure somewhat bigger than Earth forms Uranus's core, which is encased in a mass of icy slush—mostly water mixed with liquid methane, carbon monoxide and ammonia—thousands of kilometers thick. Overlying that is an equally thick, soupy atmosphere which is seven-eighths hydrogen and one-eighth helium and is laced with methane, ammonia, ethane, acetylene and ethylene. The Voyager photographs displayed the smoggy, generally featureless top layer of that greenish atmosphere.

If astronauts of some future century ever explore Uranus, they will first encounter the methane haze as they descend into the atmosphere. It will look similar to the smog that sometimes permeates the air of a large metropolis as seen from an airplane window. Eighty kilometers below that is a methane-ice-crystal fog. The temperature here is minus 215 degrees C, only 58 degrees above absolute zero. At an as-yet-undetermined level farther down, the methane cloud deck begins where the temperature rises and the

A celestial sea of moonlets, some as small as dust motes, whirls around Saturn in a thin plane that forms the spectacular rings. At this viewing angle, the Sun illuminates clouds of dust lifted out of the ring plane by electrical forces similar to static electricity. In some of the Voyager images of Saturn, the clouds are seen as "spokes" in the rings. Ring particles are mainly water ice, making them excellent reflectors of sunlight. Particles closest to Saturn orbit the planet in 5.6 hours; those at the rings' outer edge complete the circuit in 14.2 hours. This illustration is based on spacecraft findings.

Above: The 1997 discovery image of Uranus's moon Caliban (circled) was captured by the 5-meter Hale Telescope atop Palomar Mountain in California. Several images taken a few nights apart revealed the moon's motion against the background stars. Caliban is seven million kilometers from Uranus. By contrast, the nine moons of Uranus shown in the Hubble Space Telescope image at right all orbit within 35,000 kilometers of the edge of the planet's rings. The red spots on Uranus are slightly warmer regions detected by the infrared camera that took this false-color picture.

atmospheric pressure is comparable to that at the Earth's surface. The methane below that is in a liquid state, perhaps as a mist, with water-ice-crystal clouds. Deeper still, the water is liquid as well, probably as droplets suspended in the hydrogen-and-helium atmosphere. If the explorers continue down into the total blackness, they will find that the atmosphere increases in density, then turns into a slushy ocean of water, methane, acetylene, ammonia and probably liquid carbon monoxide, which extends through to the solid core. But as on Jupiter and Saturn, the surface could never be reached. Well above it, the pressure rises to levels that would crush even a heavily armored submersible like tinfoil.

Among all the planets that have been visited by spacecraft, only Uranus has a weirdly tipped "horizontal" rotation axis angled just eight degrees off the plane of the planet's orbit. The flopped axis could have been the result of a collision with a body the size of Earth during the primordial era. Uranus seems to roll around the Sun in its 84-year orbit, rather than spinning in a near-vertical position as Earth does. For stretches of time lasting up to 42 years, parts of one hemisphere are completely out of sunlight. During the 1980s and 1990s, the planet's south pole was pointed almost directly at the Sun. The amount of heat received from the Sun, however, is minuscule. To an explorer near Uranus, the Sun would appear as a tiny disk the size of a pinhead held at arm's length. Sunlight, at one-quarter of 1 percent of its intensity at Earth, would illuminate the scene to a level similar to the gloom during a thunderstorm.

Uranus's system of nine skinny rings is but a shadow of Saturn's magnificent necklace. Virtually invisible from Earth, the Uranian rings were discovered by accident in 1977 when Uranus passed in front of a star and the star's light was dimmed by the previously unknown structure. The rings are narrow and as black as soot, the opposite of Saturn's. Another difference from Saturn's rings (and Jupiter's, which are composed entirely of fine dust) is the particle size. When Voyager's radio signal was passed through the rings, analyses of the resulting distortion indicated that the rings consist of refrigerator-sized and larger boulders of (presumably) the same icy mix as the satellites.

Before leaving Uranus, a few words about its pronunciation. Most of us were taught around grade four that the seventh planet is called your-AY-nus. The preferred pronunciation, which keeps us closer to astronomy than anatomy, is YER-an-us. Astronomers use this form exclusively.

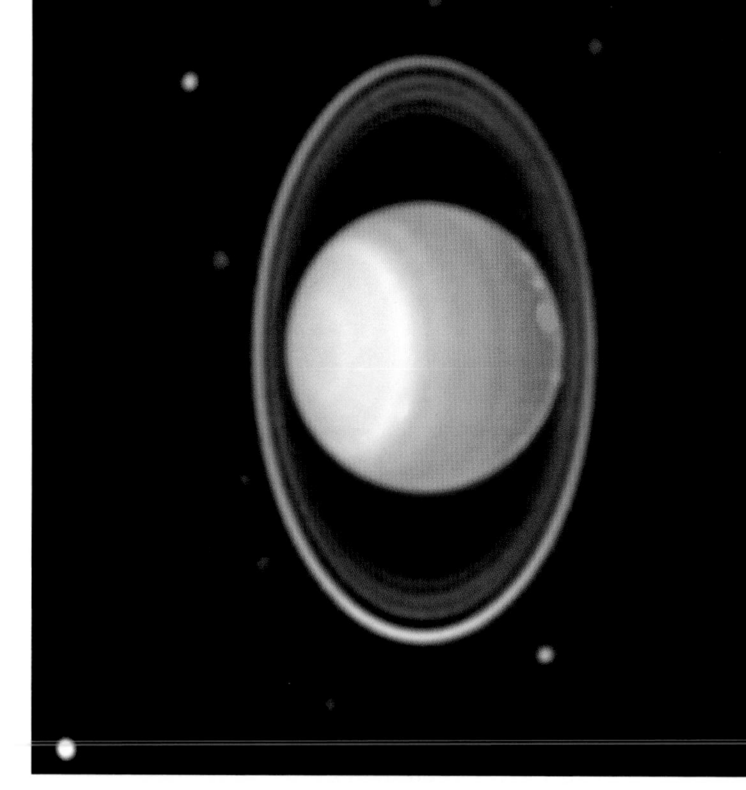

Neptune: Last of the Giants

Five months before Voyager 2's encounter with Neptune in August 1989, the spacecraft's cameras picked up a dark patch bigger than Asia floating in Neptune's atmosphere. Neptune is so remote that no specific atmospheric feature had ever been seen from Earth. But the Great Dark Spot, as it was called, proved to be an atmospheric eddy like Jupiter's Great Red Spot. Remarkably, the Dark Spot has exactly the same proportions as Jupiter's Red Spot and lies at the same latitude south of the equator. Just as Jupiter's spot is a dark shade of the creamy Jovian clouds, Neptune's spot is a dark shade of that planet's blue methane-rich atmosphere. Evidently, if the churning atmosphere of a gas giant is going to create a big spot, there is a natural size and position for it.

But unlike its Jovian namesake, Neptune's Great Dark Spot appears to be a transient feature. Six years after the Voyager flyby, the Hubble Space Telescope examined Neptune and found no trace of the huge oval. Hubble is capable of detecting even smaller features on Neptune and, in fact, has recorded some. But the Dark Spot is gone, reminding us that we still have a lot to learn about the giant planets.

Why is Neptune blue? Methane clouds and haze in its upper atmosphere absorb red light and reflect blue light. Sunlight contains both colors, so an observer looking down on Neptune sees only the blue reflected light. Uranus's aquamarine hue is also caused by methane haze, but with a slightly different mix of other compounds. Methane (natural gas) liquefies at around minus 173 degrees C and solidifies a few degrees below that. Neptune's upper atmosphere is near this temperature and shrouds the planet in a blue methane mist. A few high cirrus-type clouds of white methane crystals stream above the main blue deck.

Five rings circle Neptune well above the upper atmosphere: three thread-thin ones and two so diffuse that they could barely be detected by Voyager 2 as it hurtled by at a distance of 5,000 kilometers. Rings had been suspected because stars had flashed off as they disappeared behind them (as seen from Earth) and because all the other giant planets are decorated by ring systems. Clearly, gas-giant planets and rings are a normal arrangement.

Below the visible surface of Neptune is a deep atmosphere of hydrogen and helium that, thousands of kilometers down, eventually merges with a hot slush of hydrogen, helium and water ice sloshing around inside the planet. This core region rotates once in 16 hours, but the outer atmosphere spins more slowly and at different rates at various latitudes. The differential rotation rates create friction, which produces heat inside the planet. The heat percolates to the surface and generates turbulence—weather—which Voyager saw as the Great Dark Spot and several smaller eddies and bands.

The marked differences between the churning atmosphere of Neptune and the more featureless face of Uranus can be partly explained by an aerosol haze on Uranus that is not present on Neptune. The main difference, however, is that heat reaching the surface from within Neptune exceeds the heat received from the Sun, whereas Uranus seems to be the reverse. For comparison, Jupiter has about 50 times Neptune's internal-energy outflow and receives 20 times as much solar radiation, hence its more active cloud structure. Weak as these sources are on Neptune, there is no other way that the features we see on the blue giant could be produced.

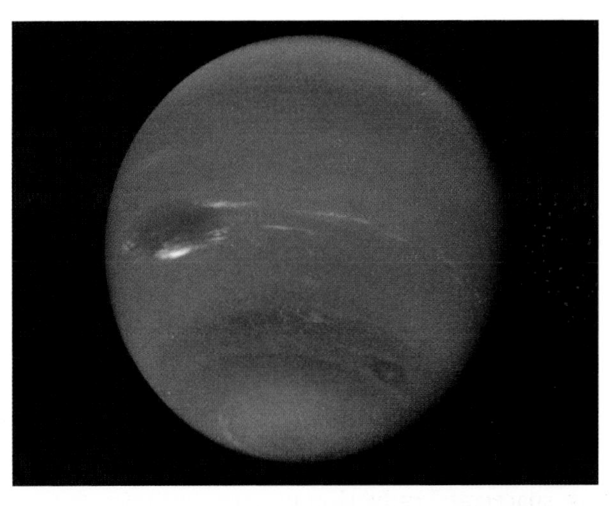

Top left: Neptune, as seen by Voyager 2 in 1989. The Asia-sized Great Dark Spot has proved to be a transient feature. It was not visible in Hubble Space Telescope images taken in the late 1990s. Left: The most detailed Voyager 2 shots of Neptune caught wispy cirrus-like clouds above the planet's atmospheric ocean of blue haze. Top: Neptune's skinny rings were captured by Voyager along with an overexposed image of the planet itself.

Pluto: Enigmatic Mini-Planet

Cruising the outer rim of the known planetary system, Pluto represents a third class of planet, different from both the terrestrial worlds and the gas giants. There is even a question as to whether Pluto should be classified as a planet at all. It is far smaller than any of the other planets and smaller than seven planetary satellites, being only two-thirds the diameter of the Earth's Moon. Many planetary astronomers define a planet as, among other things, a body with its own orbit about the Sun that does not cross the orbit of any other planet. Pluto has by far the most elliptical path of any planet in the solar system, actually sweeping inside the orbit of Neptune for two decades every orbit. Between 1979 and 1999, Pluto was the eighth planet from the Sun, Neptune the ninth. Considering how insignificant Pluto is compared with the gas giants whose realm it shares—or compared with any other planet in the solar system—it is clearly in a category of its own, which will be introduced in Chapter 4.

ICE WORLDS

CHAPTER FOUR

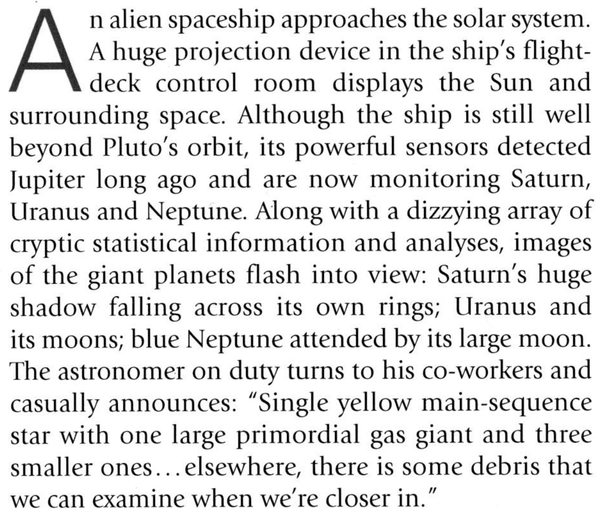

> . . . worlds unthought of until the searching mind of science laid them open to mankind.
>
> *William Wordsworth*

An alien spaceship approaches the solar system. A huge projection device in the ship's flight-deck control room displays the Sun and surrounding space. Although the ship is still well beyond Pluto's orbit, its powerful sensors detected Jupiter long ago and are now monitoring Saturn, Uranus and Neptune. Along with a dizzying array of cryptic statistical information and analyses, images of the giant planets flash into view: Saturn's huge shadow falling across its own rings; Uranus and its moons; blue Neptune attended by its large moon. The astronomer on duty turns to his co-workers and casually announces: "Single yellow main-sequence star with one large primordial gas giant and three smaller ones . . . elsewhere, there is some debris that we can examine when we're closer in."

Some debris! That's us—Earth.

Yet Earthlings frequently assess their own environment in precisely the same manner. There is more to the solar system than the Sun and its nine planets. Some of the satellites of Jupiter and Saturn are as large as the planet Mercury and, along with the rest of the moons of the giant planets, form miniature "solar" systems of striking diversity. There are oceans, volcanoes and craters in abundance. These worlds bring far more variety to the solar system than the planets alone.

The satellite systems of Jupiter, Saturn and Uranus bear more than a superficial resemblance to a scaled-down solar system—they probably formed in a similar way. The contracting clouds of material around the emerging planets made them pseudostars for a time, each with its own miniature solar nebula. In a further parallel, the billiard-table flatness of the planetary orbits is cloned in the satellite families of Jupiter, Saturn and Uranus.

Jupiter's satellite system fits into two distinct categories: very large and very small, with nothing in between. The four big moons, discovered by Galileo in 1610 shortly after the invention of the telescope, range in size from the dimensions of the Earth's Moon to those of Mercury. These Galilean satellites are worlds unto themselves.

Callisto: Battered Iceball

Outermost of the Galilean moons and the only one far enough from Jupiter's intense radiation belts to be a possible landing site for Earth explorers unprotected by cumbersome shielding, Callisto resembles the Earth's Moon or Mercury, but it's more cratered than either. Craters are jammed rampart-to-rampart everywhere on Callisto, in greater profusion than on any other body in the solar system.

A walk on Callisto would, in many respects, be like an exploration of the Earth's Moon: The two bodies have a similar surface gravity, and neither has an appreciable atmosphere. But the similarity would end upon examination of a collection of surface material. Most of the rubble and chunks of debris gathered from Callisto would melt when exposed to room temperature. Callisto's "rock" is primarily water ice with lunarlike dirt mixed in. At Callisto's surface temperature of minus 145 degrees C, water ice acquires the stiffness of rock, rather than the more plastic characteristics of glaciers on Earth. Deep in Callisto's interior, a silicate rocky body probably exists, but the bulk of the planet beyond the core is mostly water ice.

Facing page: At the frigid rim of the solar system, 30 times the Earth's distance from the Sun, Neptune and its largest moon, Triton, were first seen close up in 1989 when Voyager 2 took two portraits (here combined into a single image). The short, dark streaks on Triton are downwind deposits of frozen methane that has erupted through geyser-like vents in the moon's frozen nitrogen surface. Above: Jupiter's moon Io is seen in front of the giant planet's clouds.

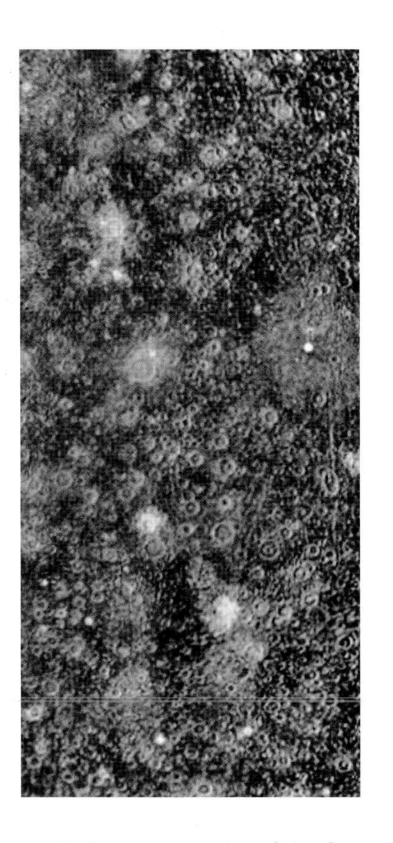

Right: A composite of the five major bodies in the Jovian system—Jupiter and, top to bottom, the big moons Io, Europa, Ganymede and Callisto. They are shown in correct size relationship to each other. For scale, Earth is three times the size of Callisto. Above: Close-up of Callisto's crater-pocked landscape. These portraits are from the Galileo spacecraft that orbited Jupiter and mapped the planet and its moons during the last half of the 1990s. This chapter presents a gallery of the best images from that mission.

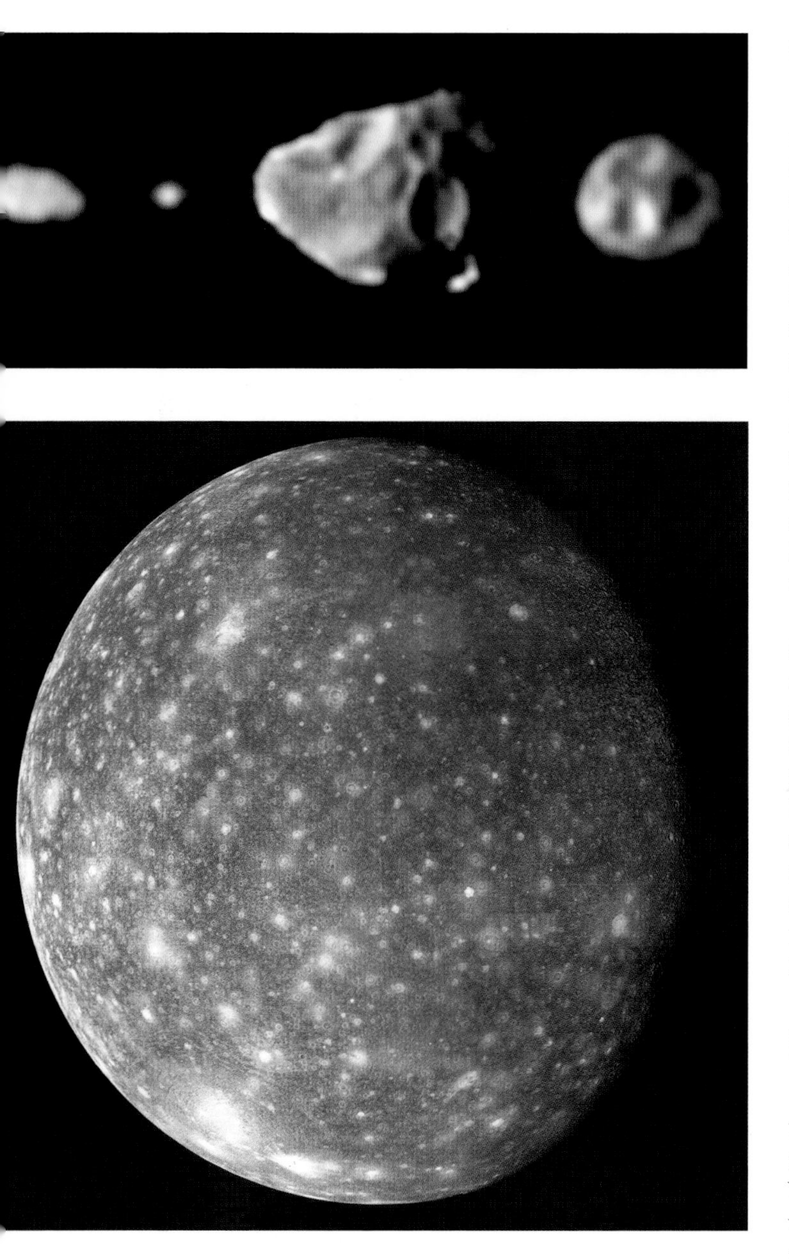

The crater-scarred surface of Callisto appears much as it must have looked four billion years ago, soon after the solar system formed. There is less evidence of change on its surface than there is on the Earth's Moon. If the other three major Jovian satellites were like Callisto, Jupiter's system would be interesting but not too much different from astronomers' expectations. However, the two Voyager spacecraft that flew past the Jovian system in 1979 and the Galileo orbiter that followed a generation later revealed that the four moons collectively represent some of the most unexpected finds in the history of astronomy.

Ganymede: Jupiter's Giant Moon

Slightly larger than the planet Mercury, Ganymede—the next Galilean satellite inward from Callisto—is the largest moon in the solar system. Like Callisto, it is a cratered world with a mainly water-ice surface darkened by dirt. Some of the impacting comets and asteroids punched through the light brown soil-ice surface composite, exposing brighter, cleaner ice underneath. The ejected icy material is splattered around these craters in what astronomers call ejecta blankets. But Ganymede has more than just craters. Large areas of ribbed and ridged icy material are spread like continents over the satellite. These regions look like glacial flows on Earth and have fewer craters than other sectors—they are clearly younger. Planetologists suggest that internal heat from Ganymede produced these features.

Ganymede is believed to be about half water ice and half carbonaceous (carbon-rich) dust and dirtlike material. Originally, it may have been fairly homogeneous internally, but radioactive elements in the soil released heat soon after the moon's formation, partly melting the ice and allowing the heavier matter to sink to the core. Ice became the predominant surface material. Apparently, Ganymede's near-planetary mass retained enough internal heat to allow a second wave of melting to form the grooved terrain. Below a thick layer of surface ice—much thicker than Europa's—there may be a slushy ocean of ice and dirt.

Europa: Ice-Encased Ocean World

Europa is the smallest of the Galilean moons and the second out from Jupiter. Until 1979, it was just another astronomy-textbook statistic. Then came the Voyager 2 close-up images, and within days, Europa was transformed—in our perception, at least—into one of the solar system's most intriguing worlds. The biggest initial surprise was the almost total lack of detail. From a distance, Europa looks like a white cue ball. At close range, the only visible features are thin, kinked brown lines resembling cracks in an eggshell. And this analogy is not far off the mark.

The surface of Europa is almost pure water ice, but a nearly complete absence of craters indicates that Europa's surface ice is thinner and less rigid than that of Ganymede and, in some ways, resembles the Earth's Antarctic icecap. The eggshell analogy may well be entirely accurate, since the ice could be as little as a few kilometers thick—a true shell around

Above: An amazing crater chain on Ganymede was probably caused by the impact of a fragmented comet similar to the 21 pieces of Comet Shoemaker-Levy 9 that smashed into Jupiter in 1994. A dozen other crater chains like this have been found on Ganymede and Callisto, indicating that comets passing near Jupiter are commonly torn into strings of fragments by the giant planet's immense gravity. Bottom left: Jupiter's moon Callisto is one of the most heavily cratered objects in the solar system. Top left: Four small, irregularly shaped moons orbit Jupiter inside the orbits of the big four. In order of increasing distance from Jupiter, they are, from left to right, Metis, Adrastea, Amalthea and Thebe. In shape and composition, they all resemble asteroids. The largest, Amalthea, is 250 kilometers long.

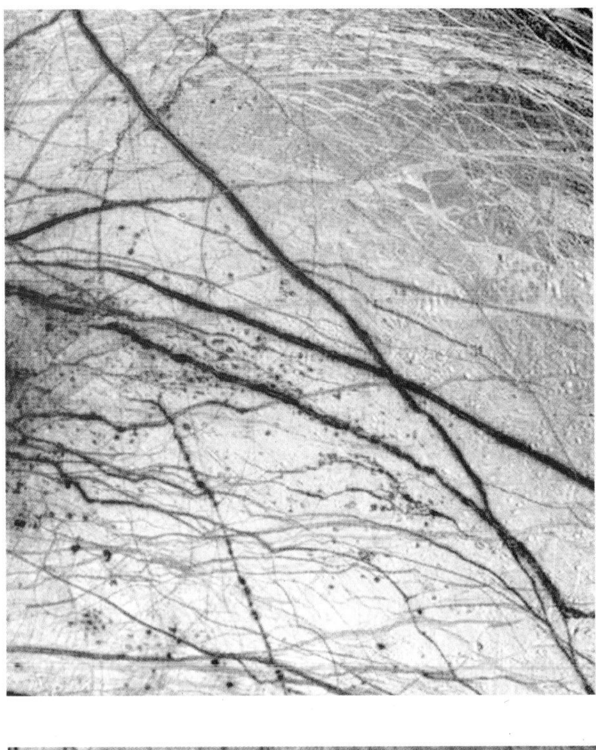

Above: For a few million years after the formation of Jupiter and its major moons, the giant planet radiated enough heat that Europa may have remained water-covered, perhaps offering life a chance to gain a toehold. Artist's concept shows primordial Jupiter with Europa in the foreground. Other images on this page are Galileo views of Europa's frozen surface today.

what is likely a subsurface ocean of liquid water that, in turn, encases a rocky core. The interior of Europa has been kept warm over the aeons by tidal forces generated by the varying gravitational tugs of the other big moons as they wheel around Jupiter. The tides on Europa are up to 60 meters high, pulling one side, then the other, then relaxing, in an endless cycle. The resulting internal heat keeps what would otherwise be ice melted almost to the surface. The cracklike marks on Europa's icy face appear to be fractures where water or slush oozes from below.

Soon after Voyager 2's encounter with Jupiter in 1979, when the best images of Europa were obtained, researchers advanced the startling idea that Europa's subsurface ocean might harbor life. Life processes could have begun near the origin of the solar system, when Jupiter was releasing a vast store of internal heat that made it more like a miniature sun than a planet. Jupiter's early heat was produced by the compression of the material forming the giant planet. Just as the Sun is far less radiant today than was the primal Sun, so the internal heat currently generated by Jupiter is but a shadow of its former intensity. During this warm phase, some 4.6 billion years ago, Europa's ocean may have been liquid right to the surface, making it a crucible for life in the same way that the Earth's ocean was. The open ocean may have lasted for only a few million years, until Jupiter began to cool.

Although just a theory, this scenario is bolstered by the discovery (which, coincidentally, occurred around the same time the Voyagers passed Jupiter) of life-forms on the Earth's ocean floor that exist in total blackness, sustained entirely by chemical rather than solar energy. Bizarre tube worms, crabs, clams and other animals and plants live around warm-water vents in deep-sea midocean rifts, relying on the sulfur and oxygen in the mineral-rich water for

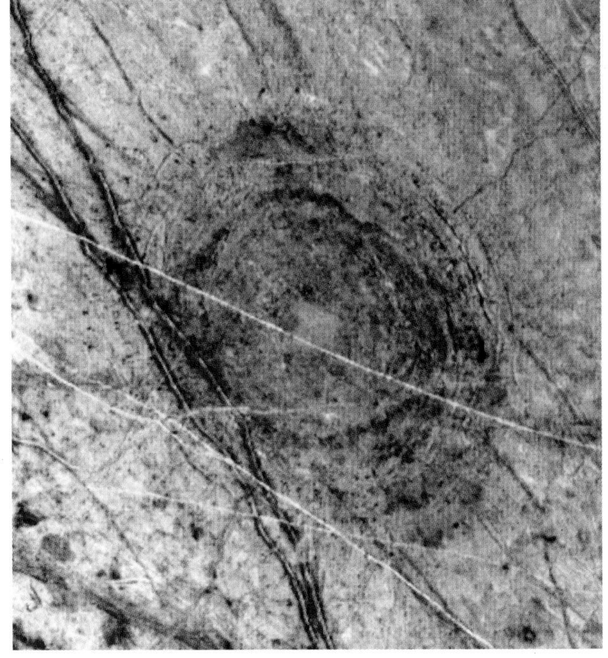

the energy required to support them. Seawater seeps through porous rock on the ocean floor, becomes heated and saturated with minerals from exposure to the hot rock below, then fountains up through chimneylike vents. As the hot water meets cold seawater, the minerals precipitate onto the surroundings,

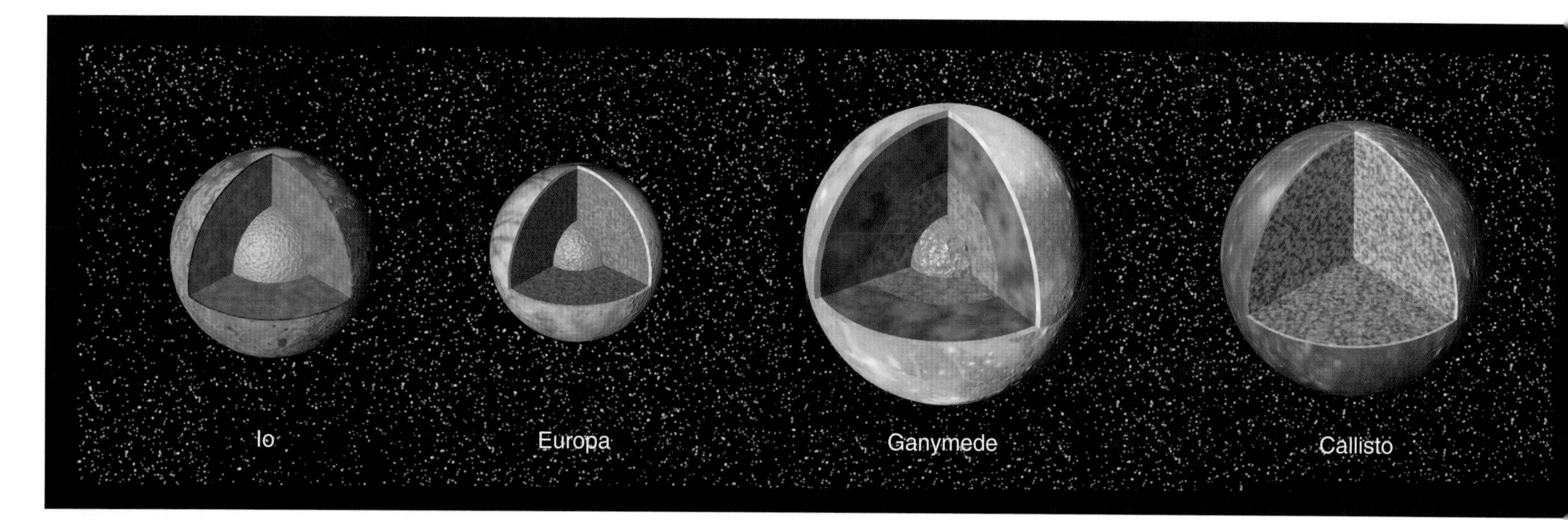

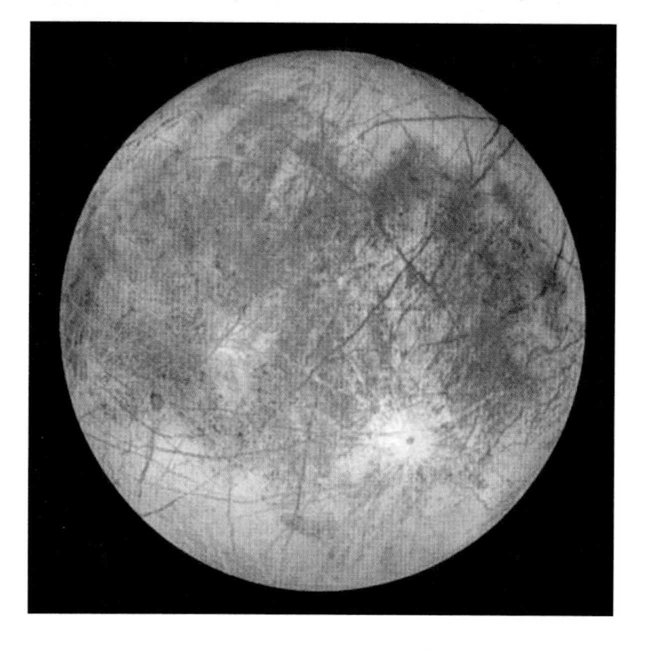

producing a fertile environment for the chemically based organisms. Until this discovery, only bacteria had been found at such depths (two kilometers) and sunlight was thought to be the sole energy source for all complex life-forms on Earth.

On Europa, the internal heat that keeps the ice-topped ocean from freezing could create the same type of ocean-bottom vents. If life began in the primal open oceans that might have existed when Jupiter was a miniature sun, the organisms could have adapted to obtaining sustenance at the seafloor by chemical means after the surface froze.

Io: Focus of Furious Forces

If Jupiter's satellite Io (EYE-oh) had ever been described in a science fiction novel, it would have been dismissed as too bizarre to be real. Here is a world where a dozen giant volcanoes are simultaneously erupting with such violence that material is ejected more than 100 kilometers into the sky. Molten sulfur oozes down the flanks of the volcanoes and spreads into the valleys below. So much matter is being expelled by the volcanic forces that this moon has literally turned itself inside out two or three times during its existence. Radiation from Jupiter's powerful magnetosphere is so intense on the surface of Io that lead shielding thicker than a bank-vault door would be barely adequate protection for a human.

About the size of our Moon, Io is the nearest of the Galilean satellites to Jupiter. As with the other big moons, Io was a virtually unknown object until Voyager 1 sped to within 20,000 kilometers of its surface on March 5, 1979. The discovery of the real Io ranks as one of the major triumphs of modern space exploration. In less than 10 days, astronomers went from thinking Io was similar to the Earth's Moon to the realization that Io is utterly alien.

As a reporter at the Jet Propulsion Laboratory in Pasadena, California—Mission Control for the Voyager probes—I witnessed the drama unfold firsthand. Voyager hurtled closer and closer to Io in early March, its electronic cameras steadily recording greater detail. But the images revealed no craters. (The more craters, the longer the object has been battered and therefore

Above: An illustration of the interiors of Jupiter's four large moons reveals that three have metal-rich cores. Callisto's interior is an amalgam of ice and rock topped by a thin icy crust. Ganymede has a thick slush layer (shown in blue) below an icy crust. Io, with its hot core and sulfurous volcanic surface, is unique among the satellites in the solar system.
Left: Full-disk view of Europa. This moon's broken-eggshell appearance and lack of impact craters indicate that its icy surface has been cracked and reshaped over time by a subsurface global ocean of water.

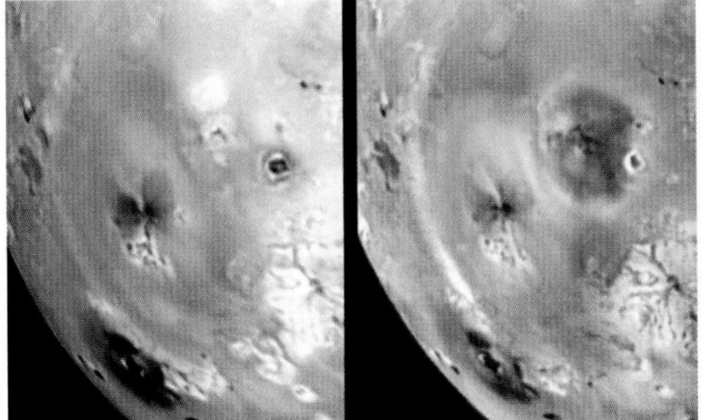

the older its surface.) The surface was a colorful quilt of white, brown and orange splotches that one Voyager scientist described as a "celestial pizza."

Three days after Voyager's closest approach to Io, Linda Morabito and Steve Synnott of the Voyager optical navigation team were comparing the position of Io with background stars, a procedure necessary for refining knowledge of the spacecraft's position and the pointing accuracy of its cameras. Using computer enhancement, they overexposed the images to bring up the background stars. Suddenly, an umbrella-shaped plume hovering above Io's surface materialized. Morabito identified it as an eruptive cloud from a volcano. Other possibilities, such as a glitch in the imaging system, were quickly ruled out, and the announcement was made: We were gazing down on a live volcano, the first ever seen on another world.

Ultimately, 12 active volcanic centers were discovered on Io, spewing enough material to blanket this tortured moon at the rate of several centimeters per century. Just as Europa's interior is heated by the oscillating tidal forces caused by the orbital interplay of the other big Jovian moons, so Io is affected. But the tides raised on Io are far greater, enough to melt its interior completely.

Io is significantly denser than the other Galilean moons because it turned its ice into steam long ago and blasted it into space through volcanic activity. Apparently, only sulfurous rocky material remains. A molten interior of sulfur and rocky lavas is overlaid by a thin crust through which the volcanoes eject their cargo. Sulfur takes on various hues at different temperatures, which is why Io's surface is so colorful. At more than 600 degrees C, sulfur is black, which is usually the color of the volcanic vent regions. As it cools, sulfur turns lighter, first brown and then bright orange, and these shades correspond with the colors of the flanks of the volcanic vents. At room tempera-

ture, sulfur is yellow, but in the extreme cold at Io's distance from the Sun, it becomes white. This is the color of the inactive plains of Io. Like exploration of the surface of Venus, an expedition to Io by humans is probably a long way in the future.

Satellite Inventory

The remaining 57 Jovian moons range from 185 to 2 kilometers in diameter—debris, compared with the Galilean satellites. The little moons can be subdivided into three groups: four inside the orbit of Io, four clustered at six times Callisto's distance from Jupiter and four at twice that distance.

Saturn's family is more diverse than Jupiter's, with one moon the size of the planet Mercury, four medium-sized moons (from one-quarter to one-half the diameter of the Earth's Moon) and 26 smaller than that. Just as at Jupiter, the Voyager spacecraft provided stunning information about Saturn's satellites. The transition from the unknown to the known was even more dramatic because for centuries, Saturn's moons, at twice Jupiter's distance, were displayed to earthbound telescopes as mere shimmering points of light. So far, Saturn's largest satellite, Titan, is the most intriguing of the group.

Titan: Glaciers and Methane Rain

In 1944, American astronomer Gerard Kuiper detected methane gas around Titan. That, plus the fact that Titan is about the size of the planet Mercury and orbits Saturn once in 16 days, was basically all that was known prior to Voyagers 1 and 2. Titan now ranks as one of the most favored sites in the solar system for further exploration and is in many ways more interesting than Saturn itself.

Titan is a featureless globe, a world cloaked in a yellowish brown smog of upper-atmosphere aerosols and hydrocarbons. Voyager experiments were able to penetrate the veil to reveal that Titan is the only place in the solar system with an atmosphere even remotely similar in composition and density to the Earth's. Our atmosphere is 78 percent nitrogen and 21 percent oxygen. Titan's atmosphere is about 95 percent nitrogen and 5 percent methane, with traces of carbon compounds. The surface atmospheric pressure on Titan is 1 1/2 times what it is on our planet. The big difference is temperature. The Earth's average surface temperature is just above the freezing point of water, while Titan's, at minus 180 degrees C, is barely above the freezing point of methane. As water on Earth exists in solid, liquid and gaseous states, so methane on Titan is solid, liquid and gas. There must be glaciers of methane ice, liquid-methane lakes and methane vapor in the lower levels of the nitrogen atmosphere that could fall as methane rain. Methane —the stuff we know as natural gas—serves the same role on Titan as water does on Earth. Ethane, a compound chemically related to methane, is likely present in large quantities too.

Humans protected by well-insulated heated spacesuits should have no problem exploring Titan. The ground would be an amalgam of methane ice and water ice that, on casual inspection, would seem identical to snow. When it snows on Titan, the flakes would hardly seem to be falling, since gravity is only one-sixth that of Earth, about the same as on the surface of the Moon. The combination of low gravity and dense atmosphere makes Titan the best place in the solar system for flying. Aircraft with small motors could easily stay aloft, and gliders would soar almost endlessly. Human-powered flight might even be possible. There are some hindrances, however, such as the need for floodlights, flashlights and helmet lamps to penetrate the gloom. By the time light from the Sun filters through Titan's dense smog and clouds, it illuminates the surface with a glow barely brighter than that of the full Moon.

Due to its vast distance from the Sun and the thickness and insulating properties of its atmosphere and clouds, Titan probably has an extremely stable weather system. The sky would appear a dull brownish color, likely featureless, without any definite in-

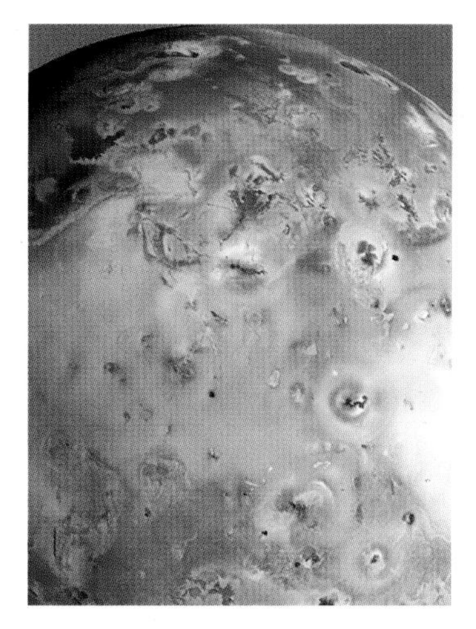

Jupiter's moon Io, above and left, is the most volcanically active body in the solar system. Typically, 5 to 10 volcanoes are erupting somewhere on this world at any one time, spewing hot sulfurous compounds up to 200 kilometers above Io's surface. At top left on this page, the Galileo spacecraft caught the plume of a volcano known as Pillan Patera. This volcano is at the center of the huge red ring at lower left in the large image on the facing page. The two smaller images on that page, taken just six months apart in 1997, show the emergence of a dark spot 400 kilometers across, created by deposits from a newly active volcanic eruption.

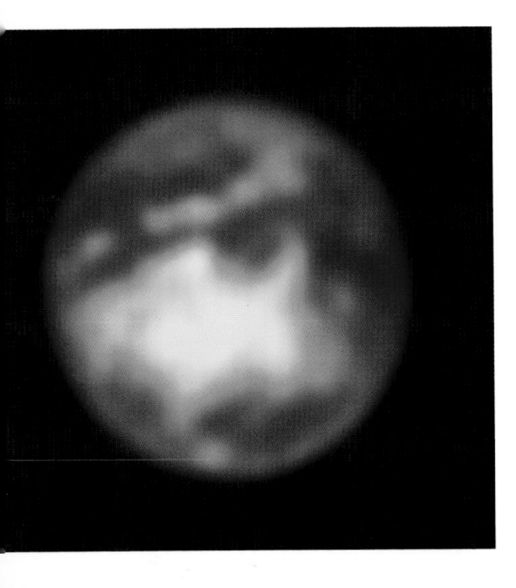

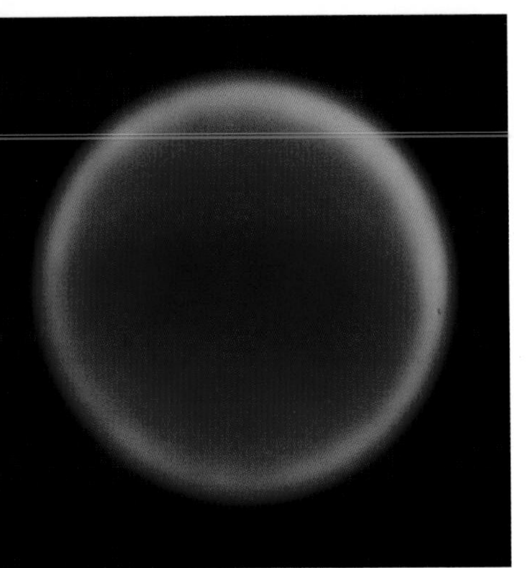

Saturn's Titan is the only planetary satellite with a substantial atmosphere (above). An infrared image from an 8-meter telescope (top) penetrates the opaque atmosphere to reveal surface features, although it is unclear what they represent. Illustration right: The Cassini spacecraft's arrival at the ringed planet on July 1, 2004. Illustration facing page: Some sectors of Saturn's moon Enceladus may still have active geysers. The moon is gravitationally torqued like Jupiter's Io, but to a lesser degree.

dication of clouds, similar to a drab winter day on Earth but less than 1 percent as bright. An expedition on the surface of Titan would resemble an exploration of Antarctica: intense cold, seemingly endless stretches of ice and snow and relatively little change in weather. And on Titan, a day is 16 Earth days long, 8 days of light and 8 days of pitch-darkness. There may be huge slush bogs. Ponds or small lakes could dot the landscape. However, according to radar readings, large oceans are doubtful. One interesting property of a methane lake is that it never freezes over—methane ice sinks, so the ice builds up from the bottom. Boats may be a 21st-century mode of transport on Titan.

Could life in any conceivable form exist in Titan's supercold environment? Back in 1961, the prolific science and science fiction writer Isaac Asimov proposed how it might happen, although he did not have Titan in mind. Trained as a chemist, Asimov realized that liquid methane will not dissolve the same kind of materials that water does, so biochemistry on Titan could not be based on proteins and nucleic acids, as it is on Earth. But methane will dissolve lipids, a class of compounds including oils and fats. On Earth, lipids form molecules whose complexity is comparable to that of protein. Compounds that develop in Titan's methane environment would remain stable, because there is no free oxygen in its atmosphere to break them down. Hence a methane biology, wrote Asimov, might produce complex organisms that would be a form of life.

Whether or not that has occurred, some of the atmospheric methane on Titan is likely being processed by chemical reactions induced by solar ultraviolet energy. The resulting hydrocarbons would rain down on the surface, leaving a layer of life precursors, perhaps as a brownish goo. One biologist has proposed that the material might accumulate in bogs up to 10 meters deep. Current theories suggest that primitive Earth, like Titan, had an oxygen-free atmosphere with significant amounts of methane, but the parallel is not complete, because the temperature on Earth was always higher. However, Titan may possess the closest parallel that scientists have found to the conditions which provided the crucible for the emergence of life on Earth.

Titan's nitrogen atmosphere was as startling a

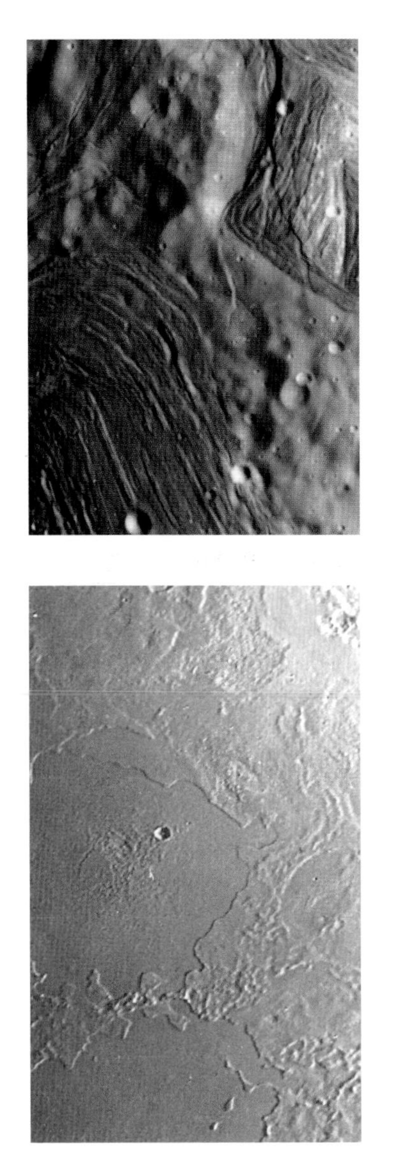

Top: A strangely grooved surface is a unique feature of Uranus's moon Miranda. Above: The plains of Neptune's moon Triton are mainly water ice frozen as hard as granite in temperatures only 40 Celsius degrees above absolute zero. Right: Saturn's satellite Dione.

discovery as were the volcanoes of Io, but in retrospect, it makes sense. Just after Titan formed, it was probably much warmer than it is now, a time when there were liquid-ammonia oceans and when gaseous ammonia was a major constituent of the atmosphere. To produce nitrogen, the ammonia was either frozen to the ground or broken into hydrogen and nitrogen by solar ultraviolet radiation. The nitrogen would then collect in the atmosphere, and the hydrogen would be lost into space. If it was ever warm enough in the early days of Titan to allow water to be liquid, life-precursor molecules could have formed in abundance. Whatever was produced must still be there, frozen in the water-ice continents.

With so many possibilities, Titan is a playground of speculation—but not for long. If all goes well, the Cassini spacecraft that arrived at Saturn in the summer of 2004 and propelled itself into orbit around the ringed planet will drop an instrumented probe into Titan's atmosphere in early 2005. Known as Huygens, after the 17th-century scientist who discovered the big moon, the probe will report on conditions both in Titan's atmosphere and on its surface.

Uranus's Satellites

Miranda, Ariel, Umbriel, Titania and Oberon—Uranus's five major satellites—all came under Voyager 2's scrutiny in 1986. Titania and Oberon are half the diameter of the Earth's Moon; Umbriel and Ariel are about one-third the Moon's diameter; and Miranda is only one-seventh its diameter. No details had ever been seen on these moons from telescopes on Earth, but Voyager's high-resolution digital camera displayed Oberon, Titania and Umbriel in detail comparable to a binocular view of our Moon from Earth. Like our Moon, all three have abundant craters.

Voyager swung close enough to Ariel to show features about the size of a small city. At that distance, some unexpected trenches and structures indicating

glacierlike flows appeared. But Miranda yielded the most surprises. Planetologists expected it to look like Saturn's moon Mimas, a crater-covered world the same size as Miranda, located a similar distance from its parent planet. Instead, Voyager revealed what one planetary scientist described as a bizarre hybrid of the geology of the planets Mercury and Mars and of some of the moons of Jupiter and Saturn. Miranda sports canyons, ribbed flow features, crevices and sectors almost devoid of craters.

At Uranus, the temperature was low enough that methane, carbon monoxide and ammonia were available in addition to water as solid moon- and planet-building materials. Water ice is like stone in the frigid realm of Uranus. But methane, ammonia and carbon-monoxide ice could flow as glaciers do here on Earth, which may explain some of the strange terrain on Miranda and Ariel.

Triton: Neptune's Companion

The triumph of Voyager 2's 12-year odyssey from Earth to Jupiter to Saturn to Uranus and, finally, to Neptune in 1989 is that the remarkable spacecraft forever transformed worlds in the remote outer solar system from mere dots, as seen in our largest Earth-based telescopes, to real places, with volcanoes, canyons and icy plains.

The last of these transformations occurred when Voyager hurtled past Neptune's largest satellite, Triton, on August 25, 1989. Two hundred close-up pictures of Triton, transmitted to Earth from Voyager's on-board image-recording system, revealed a world so frigid that water ice is as hard as granite and liquid nitrogen acts like water, forming a subsurface layer analogous to the water table beneath the Earth's surface. Natural heat from inside Triton, generated by tidal action from nearby Neptune and the decay of radioactive elements, is enough to sustain the liquid state of the subsurface nitrogen "groundwater."

Much of Triton is covered with brilliant nitrogen snow deposited from geyserlike eruptions of liquid nitrogen, which escapes to the surface through vents or cracks. Frozen methane (solid natural gas) is also released in these eruptions, leaving purple and black stains beside the vents. Dozens of elongated stains are visible in the Voyager portrait on page 54. They are elongated because of prevailing winds on Triton. Winds are not common on planetary satellites. Most, like our Moon, are airless, but Triton sports a thin atmosphere of nitrogen and methane. The atmosphere is clear except for a few wispy clouds.

Above: Pluto and its moon Charon, as seen by the Hubble Space Telescope. Only vague smudges have been mapped on these remote icy worlds. Left: NASA illustration shows the proposed Pluto Express spacecraft passing near Charon, with Pluto in the background. Although the Pluto Express was cancelled in 2000, an improved version called New Horizons, which is similar in appearance, is scheduled for launch in January 2006 for a Pluto flyby in 2015.

Triton has virtually no craters. Those which must have covered this moon billions of years ago have been erased by flooding from liquid-water "lava" perhaps two billion years ago and by more recent coverings of liquid methane and nitrogen. Although Neptune may remain off-limits to human exploration for centuries, Triton can be explored. Astronauts of the future will clamber over what scientists call the cantaloupe terrain or examine the vents thought to release melted nitrogen ice as the moon is warmed by internal heat and the change of season. Even here, 30 times the Earth's distance from the Sun, summer, however feeble, still arrives.

Future explorers will confirm or refute the idea that Triton was once an independent planet of ice and rock plying its own orbit about the Sun. The theory suggests that Triton strayed too close to Neptune and was captured by the gravity of the more massive planet about four billion years ago. For the next billion years, it swung in a highly elongated orbit, crashing into any moons Neptune originally had beyond the six known to be near the planet. Each time Triton passed close to Neptune, the tides caused by the big planet's gravity would be strongest. When Triton reached the elliptical orbit's far end, the tides would be much weaker. Astronomers calculate that the constant pulling and releasing of Triton would have generated enough internal heat to turn the moon molten. Two to three billion years ago, through natural orbital evolution, Triton's orbit became circular, the tides subsided and the surface froze. This scenario explains most of what we see, but some regions on Triton, such as the bizarre cantaloupe terrain, remain puzzling.

Planet X and Pluto

A suspected tenth planet at the edge of the solar system, dubbed Planet X by astronomers who predicted its existence throughout the 20th century, is a myth. It does not exist. The original evidence offered for Planet X was that observations of the positions of Uranus and Neptune indicated that the two planets were deviating slightly from where they should be according to orbital theory. Planet X proponents suggested that these deviations could be accounted for by the gravitational influence of a tenth planet, larger than Earth, somewhere beyond Pluto.

However, the discrepancies were the result of measurement and instrumental errors in determining the positions of the two outer planets combined with an inaccurate estimate of Neptune's mass. Once the correct mass of Neptune was obtained by Voyager 2 in 1989, researchers rechecked the old positional measures and uncovered the other errors. The locations of Uranus and Neptune as they glide along their orbits now match predictions exactly, thus eliminating any evidence of Planet X.

The search for Planet X had many parallels with the quest for the "missing planet" between Mars and Jupiter two centuries ago. Thousands of asteroids (ranging from the largest, Ceres—nearly 1,000 kilometers in diameter—to flying mountains less than one kilometer across) roam where a single planet was expected in the Mars-Jupiter gap. Similarly, a belt of comets exists where the Planet X hunters sought a single large body. The first of these objects was picked up in 1992 during a deliberate search by astronomers David Jewitt and Jane Luu, using the University of Hawaii's 2.2-meter telescope. It is a chunk of primordial ice about 200 kilometers in diameter orbiting the Sun at $1^{1}/_{2}$ times Neptune's distance.

In the years since that discovery, more than 700 similar-sized objects have been found in orbits just outside Neptune's path (30 to 50 times the Earth's distance from the Sun). Known as Trans-Neptunian

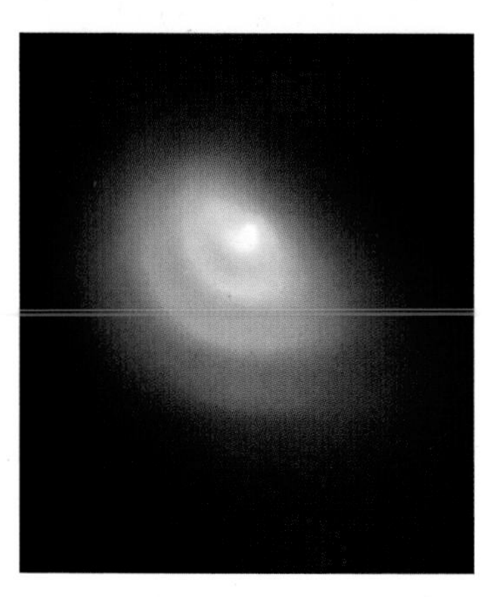

Top: The peanut-shaped 8-by-15-kilometer nucleus of Halley's Comet, unmasked in 1986 by the camera of the European Space Agency's Giotto spacecraft, shows uneven vaporization emerging from specific areas. Solar heating, absorbed by the dark surface material of the nucleus, warms the ices below the surface, producing the geyserlike activity. Illustration at right is an artist's concept of this process. The photograph above shows waves of material spun outward from erupting jets on the nucleus of Comet Hale-Bopp in 1997, a phenomenon that was visible in a small telescope.

Objects (TNOs), the giant cosmic iceballs form the Kuiper belt, named after Gerard Kuiper, the astronomer who predicted its existence in 1951. The zone could easily contain 100,000 TNOs greater than 100 kilometers across and millions of smaller ones. That's the mass equivalent of another Earth. Theorists calculate that back in the early days of the solar system, the amount of material in this region must have been much higher—enough to make another Neptune. But a fifth gas-giant planet never formed. Instead, the Kuiper belt is what remains.

In another of the many ironies in the history of astronomical discovery, we were looking at a member of the Kuiper belt all along: Pluto.

Before leaving Pluto, I must address the erroneous statement so often repeated in kids' astronomy books that Pluto roams the outer reaches of the solar system in permanent darkness, the Sun simply a bright star in the sky. Even though the Sun would appear smaller than a pinhead held at arm's length from Pluto, it still sheds light equal to about 600 full Moons. You would have no trouble reading this book by sunlight in a spaceport on Pluto.

Pluto's 248-year orbit is so elliptical that its distance from the Sun ranges between 30 and 49 astronomical units (1 AU = Earth-Sun distance), an effect which produces Pluto's seasons. The Plutonian summer occurs around perihelion, its closest point to the Sun, when the methane and nitrogen ices on Pluto

reach the sublimation point—the transition from solid to gas—to form a thin atmosphere. Perihelion occurred on September 12, 1989, so in recent years, Pluto's summer has begun to decline from its peak.

To take advantage of this season of "activity" on Pluto, a robotic flyby of the remote world, dubbed New Horizons, is scheduled for launch in 2006. From about 75 days before close encounter in 2015, the spacecraft's camera will be transmitting better images of Pluto and its moon Charon than are possible from Earth. On the day of the flyby, we should see images equivalent to telescopic views of our own Moon.

Comets

Comets are giant snowballs (water, methane, ammonia and other ices, plus a mix of dust similar in many ways to household dust), which is the best estimate of Pluto's composition. Halley's Comet, the comet we know the most about, offers some insight into the nature of these objects from the icy depths at the outskirts of the solar system. At its most distant point from the Sun, Halley's Comet curves through the Kuiper belt, but at its closest to Earth, it zooms in to Venus's distance. This long, sausage-

Comet Hale-Bopp was the largest and brightest comet visible from Earth during the last quarter of the 20th century. Its nucleus is about 35 kilometers in diameter, and its blue gas tail and white dust tail were tens of millions of kilometers long.

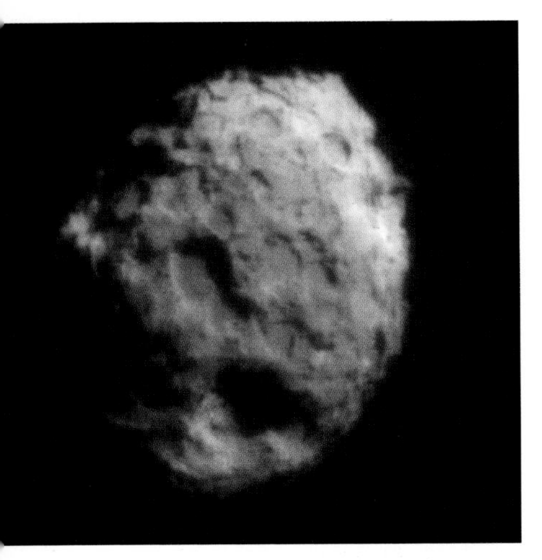

Above: NASA's Stardust spacecraft snapped this image of the six-kilometer-long frozen nucleus of Comet Wild 2 on January 2, 2004, when the comet was 390 million kilometers from Earth. Right: In March 1996, Comet Hyakutake hurtled past Earth at a distance of 15 million kilometers—close, by astronomical standards. It sported the longest tail of any comet seen in the last half of the 20th century. At its peak, the tail stretched halfway across the night sky.

shaped, 76-year elliptical orbit is typical of comets that can be observed from Earth.

When a comet nears the Sun, its surface is vaporized by solar radiation and expelled into a cloud that is swept back into the classic comet tail by the pressure of sunlight and the force of the solar wind (electrically charged particles emitted by the Sun). But it's the comets we *don't* see that interest us here. These are the ones which never get close enough to the Sun for the vaporization process to occur. These are the normal, or average, comets. Halley's Comet and all the other known comets are freaks—comets that were gravitationally disturbed from their normal abode somewhere beyond Neptune.

Halley's sweep through the Earth's skies every 76 years is the sign of a dying comet. Halley loses about a meter of ice and embedded dust from its surface during each visit to the Sun's vicinity. It can make several thousand circuits of the Sun in its present orbit, and then its store of ices will be exhausted. Eventually, all that will remain is a clod of dirt and gravelly material. But comets that stay out beyond Neptune are pristine, basically the same as they were billions of years ago. The only comets which we have detected at that distance are the larger Kuiper belt objects, but there could be as many as 100 trillion comets larger than the size of a major sporting arena, reaching out in a diffuse cloud to one light-year or more.

Triton was likely a Pluto-sized comet captured by Neptune during the early days of the solar system, when comets were more abundant. Indeed, Uranus and Neptune are themselves related to comets. Their compositions of water, methane, ammonia and hydrogen are in the same approximate ratio as those elements and compounds in the ices of comets.

During the solar system's formation, trillions and trillions of comets condensed from the solar nebula in the region where the outer planets formed. These were the planetesimals of the outer solar system, the building blocks of the giant planets. Once Uranus and Neptune were established, they became the gravitationally dominant bodies. Like tag-team wrestlers, they soon threw the rest of the comets out of the ring. Some were flung completely outside the solar system, while the rest were hurled from their nursery to form the Oort cloud, beyond the influence of the planets. Although the Oort cloud has never been observed directly, it must exist, to account for the comets that

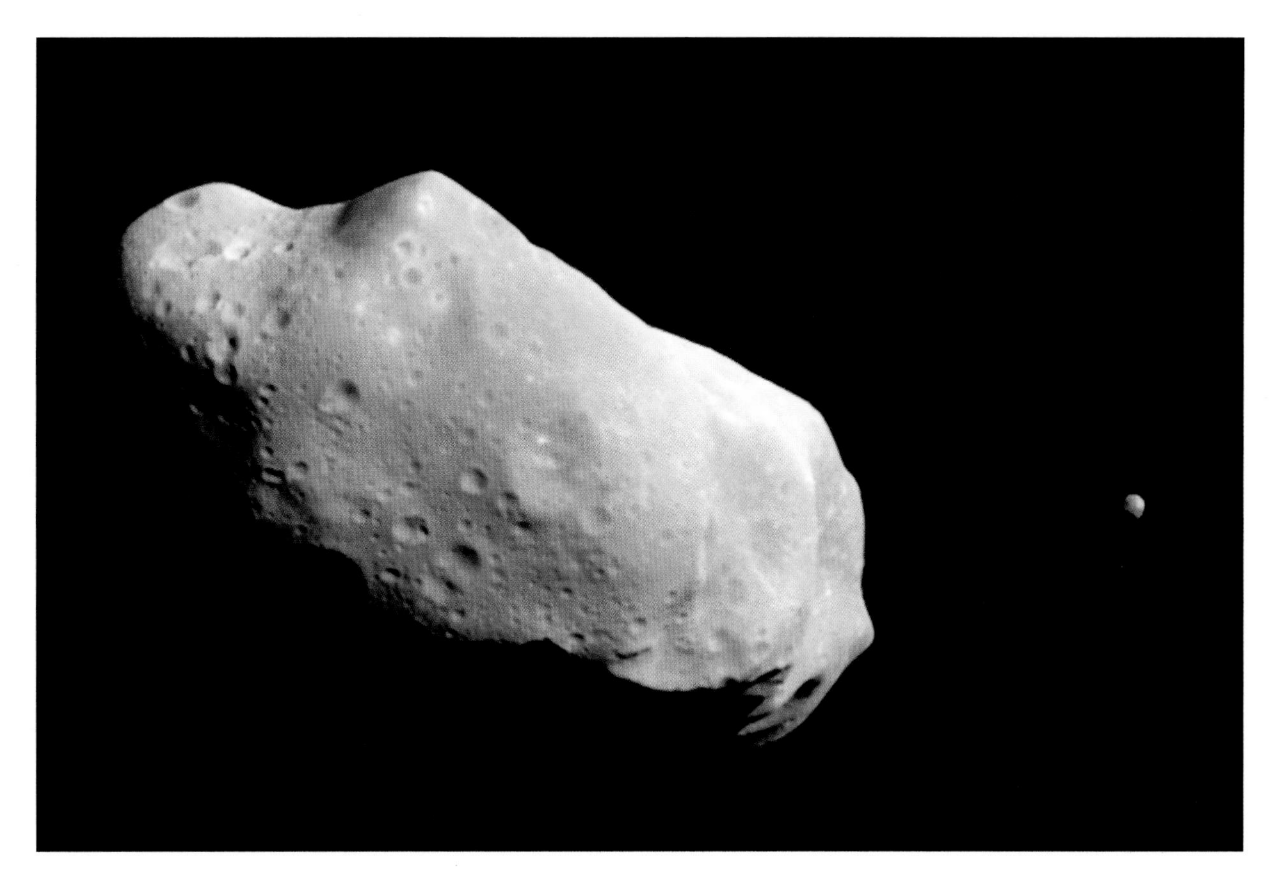

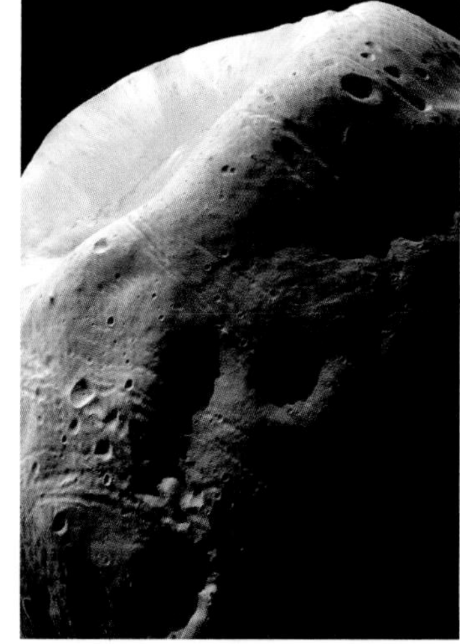

Left: Asteroid Ida and its tiny attendant moon Dactyl. On its way to Jupiter in 1993, the Galileo spacecraft was targeted to fly past the asteroid Ida. In the process, Galileo uncovered a moon circling the asteroid. Both are seen in this remarkable portrait. Ida is 56 kilometers long; its moon is less than 2 kilometers wide. Above: The Martian moon Phobos is almost certainly a captured asteroid.

arrive in our vicinity from enormous distances. The Kuiper belt is believed to contain the comets that remained roughly where they were born and thus survived the mayhem nearer Neptune and Uranus.

The normal long-term motion of nearby stars, which sometimes brings them within one or two light-years of the Sun, combined with the gradual shifting of mass distribution in the arms of the Milky Way Galaxy around us, occasionally provides enough gravitational deflection to disperse some of the comets from their remote outposts and dump them toward the inner solar system. That is how Halley's Comet and the other visible comets arrived. Comet Hale-Bopp of 1997, one of the brightest comets in recorded history, came in from 15 times Neptune's distance, tracing a huge 4,200-year orbit around the Sun.

Asteroids

More than a million asteroids populate the zone between Mars and Jupiter. Ceres, the largest, is about the same width as France; the smallest asteroids are mere flying boulders. A few renegades roam inside the orbit of Mars or beyond Jupiter, but the bulk of them have stable orbits in the Mars-Jupiter gap. When we look at a plan of the solar system, we see this as a real gap. Not only does it separate the giant planets from the smaller terrestrial worlds, but it's large enough to accommodate a planet easily. Yet none formed there. The prevailing theory is that Jupiter's gravitational influence so perturbed this region of the solar nebula that one large planet was never able to develop. Instead, several smaller ones did, some of which collided, and later the fragments collided, and so on, resulting in the range of sizes we see today.

The total mass of the asteroids is probably less than that of the Moon. In composition, asteroids range from chunks of nearly pure nickel/iron to carbonaceous bodies similar in some ways to garden dirt. Ceres is one of the primordial mini-planets. There may be others, but most of the asteroids are fragments of those original bodies. In 1991, on its way to Jupiter, the Galileo spacecraft obtained the first close-up look at an asteroid. Gaspra is 20 kilometers long and 11 kilometers wide, approximately the same size as the two moons of Mars, which themselves are almost certainly captured asteroids.

Of most interest to Earthlings are the several

Right: View toward the Sun from the surface of Sedna, the most distant known planetoid. A portion of the Milky Way Galaxy is seen at left in this illustration. The elongated glow around the Sun is sunlight illuminating the fine dust in the plane of the solar system. Below right: The colossal orbit of Sedna dwarfs the orbits of Neptune and Pluto. The Kuiper belt (not shown) is just outside Neptune's orbit but inside the orbit of Sedna. Above: Sedna's orbit is shown here relative to the Oort cloud (blue), the solar system's largest reservoir of comets. Sedna does not seem to be part of the Oort cloud or the Kuiper belt.

dozen asteroids that approach closer to Earth than either Mars or Venus. The largest, named Eros, is a brick-shaped chunk of asteroidal material three times as long as it is wide, about the dimensions of Manhattan Island. It cruises three times closer to Earth than does Mars. Smaller asteroids less than a kilometer across occasionally glide past at only a few times the Moon's distance. These flying mountains are visible for a few days as faint specks of light traversing the remote star field. Searches turn up a couple of new Earth-approachers every year, indicating that there must be hundreds. All of them originate in the main asteroid belt. Gravitationally nudged by a close encounter with a companion, they swing our way. Statistically, such an object clobbers Earth every million years or so, exploding on impact with a force far exceeding the power of the largest hydrogen bomb.

Sedna: Farther Out Than Pluto

In February 2004, a mysterious planetlike body three-quarters the size of Pluto was discovered orbiting the Sun in a huge elliptical path well beyond Pluto's orbit. Provisionally named Sedna (the "goddess of the ocean" in Inuit mythology), the object is currently 13 billion kilometers distant, deep in the remotest reaches of the solar system.

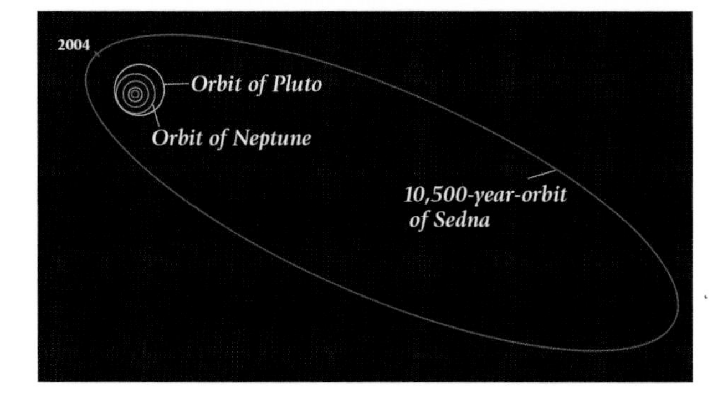

Nothing as big as Sedna has been found in the solar system since the discovery of Pluto in 1930. Some astronomers have suggested that Sedna may be the first detection of one of the largest objects in the Oort cloud, the faraway repository of billions of small, icy bodies that supplies the comets which occasionally cruise through the inner solar system.

Sedna's 10,500-year elliptical orbit of the Sun is more cometlike than planetlike. At its most distant,

Sedna is 130 billion kilometers from the Sun—30 times Pluto's distance and 900 times the Earth-Sun distance. Sedna will gradually move closer to the Sun until 2075, when it begins its multimillennial return trip to the far reaches of the solar system. The last time Sedna was this close to the Sun, Earth was just coming out of the last ice age.

Sedna's elliptical orbit resembles that of objects predicted to lie within the Oort cloud, which surrounds the Sun and extends outward almost halfway to the nearest stars. But Sedna doesn't exactly fit the parameters, because it is several times closer than the predicted inner boundary of the Oort cloud. Sedna is thus an enigma, appearing to have an orbit marching through a region too far out for the Kuiper belt but too close to the Sun for the Oort cloud.

However, the Oort cloud may have been disturbed long ago by the gravity of a rogue star that passed near the Sun in the solar system's early days. Sedna may be evidence that such an encounter occurred, making the inner Oort cloud closer than we think. But for the present, where Sedna fits into the solar system's history is highly speculative.

Soon after its discovery, researchers turned the Hubble Space Telescope toward Sedna, hoping to find evidence of a moon, which in turn would allow a more accurate value for Sedna's mass. But no moon was detected.

In addition to the intrinsic surprise of its mere existence, the discovery of Sedna has brought into focus the debate over the definition of a planet. Ever since the mid-1990s, when astronomers realized that Pluto may be the largest member of the Kuiper belt, some researchers have been asking the International Astronomical Union (IAU) to draw up an official definition of a planet. Pluto would fail many of the proposed criteria. When Pluto was discovered occupying that apparent void beyond Neptune, nothing was known or suspected of the Kuiper belt, so it seemed logical to call it a planet. In light of recent discoveries, however, many astronomers think Pluto should be demoted.

"There is a lot more evidence now that it really was a mistake to call Pluto a planet," comments Brian Marsden, head of the IAU's celestial nomenclature committee headquartered at the Harvard-Smithsonian Center for Astrophysics in Cambridge, Mass. Others argue that the three-quarter-century tradition of a nine-planet solar system should be retained, regardless of the historical quirk of Pluto.

Though the definition of a planet—and the status of Pluto—has become a tense debate in astronomy over the past decade, nothing has energized the discussion more than the discovery of Sedna. Some astronomers say that if Pluto is a planet, Sedna is clearly a planet too. Almost everyone concedes, though, that it's time to end the confusion and agree once and for all on a precise definition.

One possible definition: A planet is any solid object that is orbiting a star and is more than 1,000 kilometers in diameter—the diameter of Ceres, the largest asteroid in our solar system. That would mean both Pluto and Sedna would qualify. Ask five astronomers for a definition, however, and more than likely, you will get five different descriptions, all agreeing in general but varying in detail.

In March 2004, a committee of the IAU made up of a dozen scientists began discussions on how to settle the issue. "There is some urgency now," says astronomer Iwan P. Williams, a professor at University of London Queen Mary and Westfield College, who will chair the group. According to Williams, the committee is planning to make a recommendation in 2005, but no formal approval will be given until the IAU's general assembly in 2006.

This illustration of the comparative sizes of some of the smaller bodies in the solar system places Pluto in perspective as a world smaller than the Earth's Moon. Measuring three-quarters the diameter of the 2,340-kilometer-wide Pluto, Sedna is the largest-known body orbiting the Sun that is not called a planet. Quaoar, the biggest Kuiper belt object, was discovered in 2002.

COSMIC FURNACES

CHAPTER FIVE

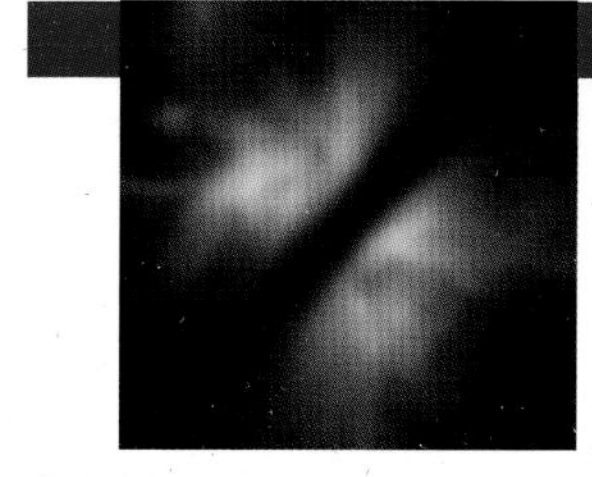

How is it that the sky feeds the stars?

Lucretius

There was a time when talking about other suns and planets could be dangerous to one's health. In 1600, Italian philosopher Giordano Bruno was burned at the stake partly because he maintained that "countless suns" and "an infinity of worlds" exist, a heretical notion at the time.

Stars are suns, of course, and there are more of them in the universe than there are grains of sand on all the beaches on Earth. They are the citizens of the galaxies. Like people, some stars are young, some old, some big, some small, some more noticed than others. If we could take a census of the Milky Way Galaxy's hundreds of billions of stars, the Sun would come out quite respectably in the top 5 percent, larger and brighter than average. It would be further distinguished because our Sun does not have a companion star. Two-thirds of the stars in our sector of the galaxy are multiple-star systems, usually binaries, but sometimes triple and quadruple groupings exist in a complex interweaving of gravitational forces. Rare, though by no means unknown, are families of five or six stars, all bound together by gravity's grip.

These multiple-star systems are not analogues of solar systems, however. They arrange themselves in a different way because the objects are closer in mass to each other, unlike our solar system, where the Sun outweighs the largest planet by a ratio of 1,000 to 1. A case in point is our nearest stellar neighbor, Alpha Centauri, a triple system 4.4 light-years away led by a star much like our Sun. The three stars in the system are known simply as Alpha Centauri A, B and C. Alpha Centauri B orbits A at a distance approximately equal to that between the Sun and Uranus. Alpha Centauri C—also known as Proxima Centauri because it is two-tenths of a light-year nearer to our solar system and is technically the closest of all stars to us—has a gigantic million-year orbit around A and B. In a gravitational sense, Proxima Centauri feels the combined masses of A and B as a single point. (A somewhat similar situation exists in our own solar system: Earth and the Moon share a common center of gravity which is closer to the Earth's surface than its center, and that gravitational focus, rather than the Earth's center, is the point orbiting around the Sun.)

Alpha Centauri A is slightly more massive than our Sun and about 50 percent brighter. Its nearest companion, Alpha Centauri B, has about 90 percent of our Sun's mass but only 40 percent of its luminosity. If a planet were orbiting Alpha Centauri A at the same distance that Earth orbits the Sun, star A would appear the same size as our Sun but would feel distinctly warmer. Life might be clustered closer to the planet's poles. More likely, though, an Earth-sized world at this distance from such a star would have surface conditions like those on Venus.

Alpha Centauri A and B revolve in elliptical 80-year orbits around a common center of gravity. When the stars are closest, they are a little more than 10 times the Earth-Sun distance from each other. At their farthest, they are three times that distance. From our hypothetical planet in orbit around A, star B would look like a dazzling point of light most of the time. At its nearest, it would be a small, brilliant disk. Whenever B was in the night sky, the landscape of star A's planet would be illuminated to twilight level and only the brightest of the more distant stars would be visible to the unaided eye. One of those—our Sun—would appear as

Splashed across the backdrop of the southern constellation Vela are the tattered remains of a star that exploded about 9,000 years ago. This is the nearest and most spectacular example of a supernova remnant. The explosion must have illuminated the Earth's night sky as brightly as a nearly full Moon. Yet all that is left of the star are these filaments and a dense cinder, called a neutron star, no wider than a small city. As the shock waves and gas from a supernova explosion push outward into space, they may meet a nebula —a cloud of gas and dust—and trigger its collapse, creating new stars and renewing the cycle of birth and death in the cosmos.

The planets and moons in our solar system are a varied lot, but they don't include the full range of possibilities. This illustration offers a more complete inventory of hypothetical worlds, especially those more massive than Earth but less massive than Uranus and Neptune, which represent a big gap in our solar system. Another gap is planets between the sizes of Mars and Earth. If Mars were half the Earth's mass instead of one-tenth, the red planet would be even more interesting than it already is. The real universe, with its seemingly endless array of stars, undoubtedly harbors worlds we cannot begin to imagine.

a first-magnitude star in the constellation Cassiopeia.

The third star in the system, Proxima Centauri, is extremely faint and is 12,000 times the Earth-Sun distance from the A and B pair. From a planet orbiting A, Proxima Centauri would be barely visible to the unaided eye, a dim star that, if it were orbiting our Sun at the same distance, might not have been noticed until after the invention of the telescope.

Could a planet like Earth exist in the Alpha Centauri system? For many years, this scenario was considered highly unlikely, since star B's varying distance from A would set up oscillating gravitational torques on the orbit of an Earthlike planet that would, over time, drastically modify its path around A. But these assessments have been reexamined, and it is now thought that a stable orbit may be possible for millions of years. But whether the orbit would remain stable for *billions* of years, as in our solar system, is uncertain. The question is important because of the abundance of multiple-star systems. Some experts suggest that multiple-star systems may not be born with planets in the first place because of the differ-

Illustration left: In comparison with our Sun, the red-dwarf star Proxima Centauri is a weakling that sheds only a feeble glow onto a hypothetical planet and its companion moon. Because they are known emitters of flares proportionally larger than the largest eruptions on our Sun (see illustration, page 77), red dwarfs are considered unlikely to have habitable planets. Illustration above: Dust particles coated with water, methanol and carbon dioxide ices—fundamental ingredients for life—have been detected in abundance in planetary "construction zones," such as the newly formed planet (foreground) of a youthful Sunlike star in this rendering.

ent speeds and motions of the nebular gas and dust during the birth of two or more stars close together. Pairs of stars separated by more than 10 astronomical units (1 AU = Earth-Sun distance) may be the only ones capable of having planetary families.

Multiple stars are systems within systems, but they are always arranged in increments of ones and twos: a single star orbiting around a pair, a pair of stars orbiting a single star, two pairs of stars orbiting about a common center of gravity or, the usual situation, simply a pair of stars. The two stars in a binary system can be almost in contact with each other or up to a light-year apart. Of course, there are billions of single stars like the Sun, all of which could have planets.

Planets of Other Suns

Astronomers and philosophers have debated the question of life on other worlds for centuries. The pendulum of opinion on the subject swung to its greatest extreme in the 18th century, when many astronomers were convinced that every star must have its attendant planets and that all planets were inhabited. William Herschel, who discovered the planet Uranus in 1781, even speculated that the Sun was populated. He thought that the dark sunspots were openings in the Sun's fiery atmosphere through which we might peer into the habitable realm below.

In the early 20th century, opinion reverted to the view that we may be alone in the cosmos. The Sun's family of planets was regarded as a fluke of nature, wrenched out of the Sun's gaseous body by the gravitational influence of a star that sideswiped the Sun. Since stars are so far apart, such an encounter would be exceedingly rare. If our planetary system had been created in that way, it would probably be the only one in the Milky Way Galaxy.

By the middle of the 20th century, that hypothesis was discarded for a variety of theoretical reasons, the main one being that material torn from the Sun would be far more likely to disperse into space or to fall back into the Sun than to coalesce into planetary bodies. Today, both theory and observation support the idea that stars are born from nebulas—clouds of dust and gas—and that planets form from abundant leftover material spread in a disk around the newborn sun. Planet birth *seems* to be a natural by-product of star birth. The concept would get a big boost if a planetary system resembling our own could be detected orbiting another star. But direct observation

Some of the giant planets discovered orbiting Sunlike stars are several times Jupiter's mass, yet they orbit their parent suns at roughly the same distance that Earth orbits the Sun. Such worlds could easily boast Earth-sized moons, as depicted in the hypothetical scene above.

of planets of other suns remains out of reach. Even the Hubble Space Telescope has not spotted a planet as large as Jupiter. The difficulty is not so much the distance to other stars but, rather, the overwhelming brightness of a star compared with its planets. It is like trying to distinguish a penlight beside a searchlight. Nevertheless, a clever *indirect* technique has succeeded in measuring the mass and orbital period of Jupiter-sized planets circling a growing list of other stars (see page 173).

The idea is to search for tiny but measurable shifts in the positions of absorption and emission lines in the spectrum of hundreds of Sunlike stars. Each spectral line has a characteristic position depending on the element that produces it. The lines shift collectively toward the red end of the spectrum if the star accelerates away from us and toward the blue end if its motion toward us increases. The changes one way or another from an average value are of interest to researchers, because such shifts reflect gravitational tugging to and fro as a large planet orbits the target star. Astonishingly, these variations amount to barely more than walking speed and were simply too small to measure until the 1990s, when technology caught up with theory. But persistence has paid off handsomely, and more than 100 planets ranging from about half of Saturn's mass to many times Jupiter's mass have been captured. Experts now unanimously agree that the spectral oscillations do indeed represent the gravitational signature of orbiting planets and not some other effect, such as changes in the stars themselves.

One of the most intriguing of these newfound planets of other stars is the trio of worlds discovered in 1999 orbiting Upsilon Andromedae, a star 47 light-years away that is slightly more luminous and more massive than the Sun. The inner planet is about three-quarters the mass of Jupiter and orbits at one-fifth Mercury's distance from the Sun. The second world is at least twice Jupiter's mass and is a bit farther out than Venus's distance. The third member of the system is four times Jupiter's mass and circles at half of Jupiter's distance. The spectrum-analysis technique does not, as yet, allow detection of planets less massive than Neptune, so we may be decades away from bagging an Earth-mass body. But considering the discoveries so far—especially the fascinating Upsilon Andromedae system—is it reasonable to assume that if giant planets are this abundant around a single star, surely Earth-sized worlds must be abundant as well? Not necessarily.

As a star and its planets are born, theorists say that the full range of bodies, from small rocky chunks near the star to hulking gas giants farther out, would emerge from the primordial nebula. But suppose that *after* the planets are formed, the giant planets move in closer to the star. This seems to have happened to the vast majority of the planetary systems discovered so far. In the Upsilon Andromedae case, no Earth-sized worlds at anywhere near the Earth's orbital position could possibly survive for long with such dangerous neighbors. We would have been gravitationally hurled

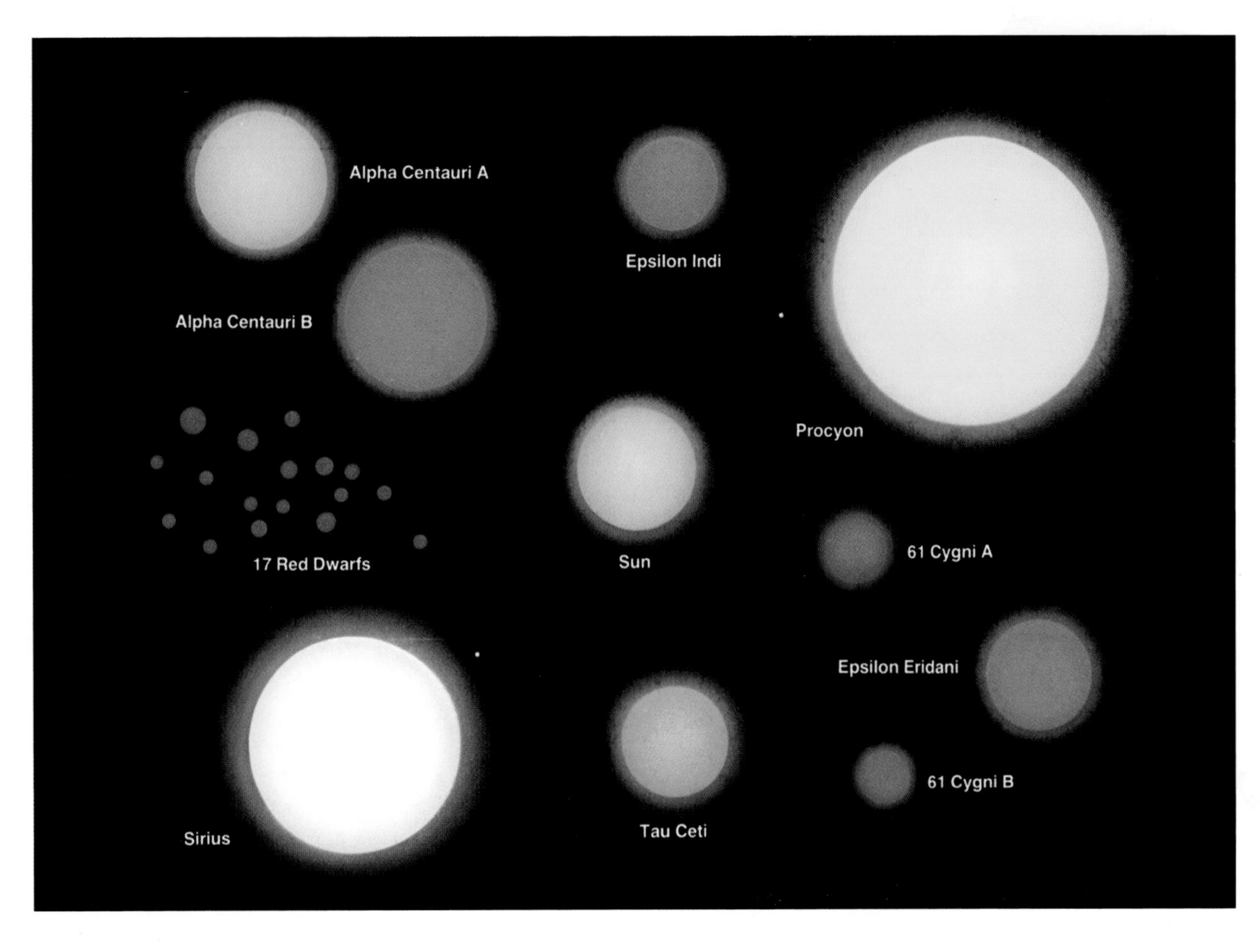

into deep space billions of years ago. The question that remains is, Are we normal or rare? Is a planetary system with rocky inner planets and gaseous/icy worlds farther out ordinary or extraordinary? At present, it is much easier to detect a massive planet that orbits close to its parent star than one that orbits at Jupiter's distance or farther, so it is unclear what's "normal" and what isn't.

Is It a Star or a Planet?

Stars and planets are part of a hierarchy of the cosmos that includes a vast range of objects differentiated by their masses. Our solar system provides an inventory which demonstrates a portion of that gradation. At one end of the scale is the Sun, a star over a million times the Earth's volume and 333,000 times its mass. At the other end are cosmic dust particles, visible only under a microscope, that constantly rain into the Earth's atmosphere at the rate of hundreds of tons a day (the silt in a house's eaves probably contains a minute amount of interplanetary material). Between these two extremes are worlds ranging from Pluto, less than 1 percent of the Earth's mass, to Jupiter, 318 times as massive as our planet. Overlapping the lower end of the regime are the planetary satellites, with Jupiter's Ganymede, at $1^1/_2$ times the Moon's diameter, leading the pack that extends down to the smallest planetary moons, a few kilometers in diameter. Below this size are the vast numbers of asteroids and comets. The major gap in the otherwise continuous size gradation in our solar system is the thousandfold increase in mass from Jupiter to the Sun.

The Sun is often referred to as a typical star, but actually, the average star in the galaxy is considerably smaller and dimmer. Astronomers call these plebeian stars red dwarfs, because of their tint and small size. Red dwarfs constitute more than half of the population of the entire Milky Way Galaxy. There are at least 100 billion of them, yet not a single one is visible without a telescope. Proxima Centauri, the nearest star, is a red dwarf. It is about one-tenth the diameter of our Sun and one-tenth its mass, but it

Left: Comparative sizes and colors of all stars within 12 light-years of the Sun. Red dwarfs are the distinct majority. The smallest stars are white-dwarf companions to Sirius and Procyon. Above: The largest planet possible is a gas giant about three times Jupiter's mass. More massive planets are compacted by their own weight to Jupiter's size or slightly smaller. Shown below Jupiter, to scale, are Neptune, Earth, a red dwarf and the Sun.

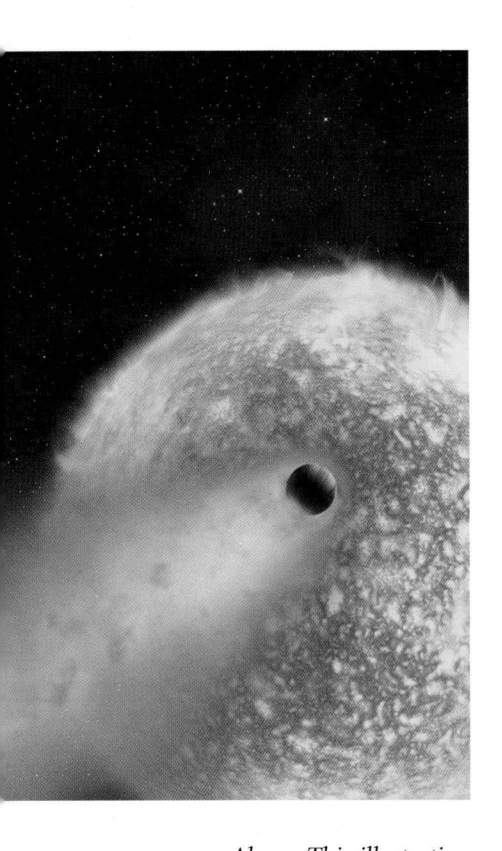

Above: This illustration depicts a Jupiter-sized planet orbiting only seven million kilometers from the Sunlike star HD209458. This is so close, the planet is evaporating from the intense heat and radiation. The system was discovered in 1999 when the star displayed a slight drop in brightness as the planet moved in front of it. Right: Eta Carinae, located at the heart of the largest nebula in our sector of the Milky Way Galaxy (photo, page 82), is one of the most luminous and massive stars known. In 1843, it brightened a hundredfold, ejecting this huge cloud of gas into space. Sooner or later, a similar eruption will occur again.

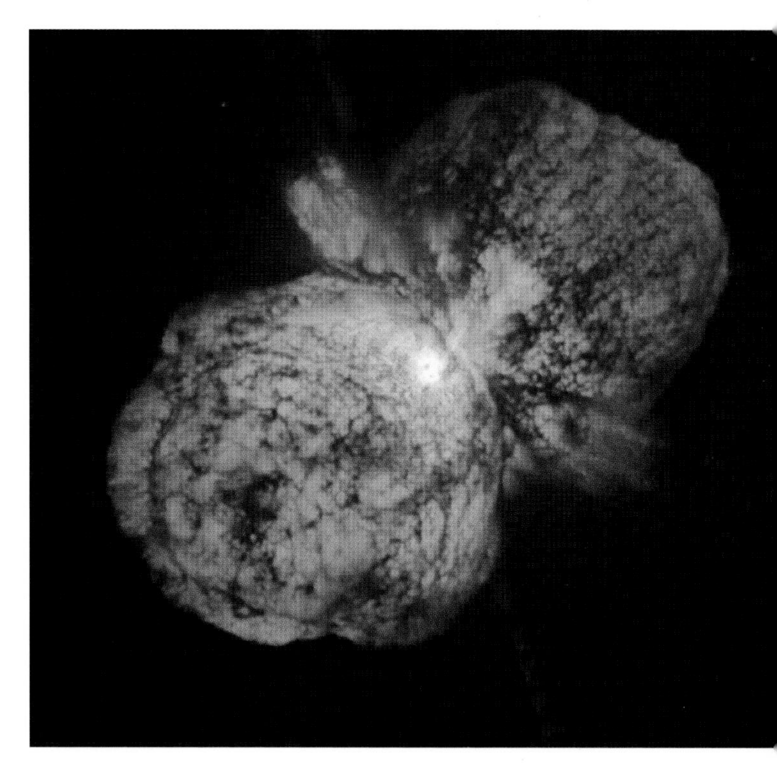

shines with only one twenty-thousandth the brightness. If Proxima Centauri were to replace the Sun, daytime on Earth would never be more than a deep and somber twilight.

The faintest red-dwarf stars known are less than one-millionth the brightness of the Sun and only detectable in large observatory telescopes out to a distance of a few dozen light-years. At just 8 percent of the Sun's mass (80 times Jupiter's mass), these stars are near the borderline between true stars like the Sun, which produce energy by fusing hydrogen into helium, and the less massive brown dwarfs, which cannot ignite the thermonuclear furnace and merely generate a dull glow through compression—the mechanism responsible for Jupiter's significant internal heat.

Brown dwarfs just below the 80-Jupiter-mass transition line masquerade as red dwarfs for about 100 million years by releasing significant amounts of energy provided by gravitational contraction. But once they compress to 90 percent of the diameter of Jupiter, they stop shrinking and cool off, slowly dimming over the next few billion years. Predictably, the less massive a brown dwarf is, the more difficult it is to detect. The first one wasn't unambiguously identified until 1995.

One of the most interesting brown dwarfs discovered so far is part of a binary star known by the awkward catalog number 2MASSW J0746425+2000321. The brown-dwarf component is 6.6 percent of the Sun's mass and is paired with a red dwarf at the minimum red-dwarf mass of 8 percent. Masses of stars in a binary system can be computed with much higher precision than those of lone stars.

If a brown dwarf of roughly this size, say, 60 Jovian masses, were to replace Jupiter in our solar system, the changes would be surprisingly minor. The great mass might disturb Saturn's orbit, but the rest of the planets would ply their paths around the Sun essentially unaffected. Since Jupiter and a brown dwarf of this mass are nearly the same diameter, the brown dwarf would appear as Jupiter does in the Earth's sky—merely as a bright "star"—but a little more brilliant than Jupiter, because its own internal light from compression would be added to reflected sunlight, which would give it an orange tint. It would probably have dramatically active cloud circulation, possibly like Jupiter's but more rapidly modified by currents welling up from hotter regions below.

Just as Jupiter has the largest satellite system (by mass), so we would expect the more massive brown dwarf to have an even bigger retinue. There is no reason a moon of a brown dwarf could not be as large as Earth. Such a supermoon, in an orbit about the size of Europa's, would be heated to nearly room temperature on the inward-facing side. (All major planetary moons become gravitationally locked from youth, so one side perpetually faces the parent body.) That might provide an opportunity for life to emerge, though the specific environment is difficult to predict, since we have no experience with a world permanently warm on one side and frigid on the other. Brown dwarfs are probably as common as red dwarfs; they are simply too dim to be identified. Earth-sized objects orbiting brown dwarfs may be rare, but once found, they should be at least as interesting to visit as Europa or Titan.

Sun Death

The Milky Way Galaxy is the birthplace and graveyard of trillions of stars. Ubiquitous red dwarfs have the greatest life spans because they cook their hydrogen fuel slowly at low temperatures. A red dwarf like Proxima Centauri will radiate at its current rate for at least 200 billion years—more than 10 times longer than the present age of the universe. At the opposite extreme are blue supergiants such as Rigel and Deneb.

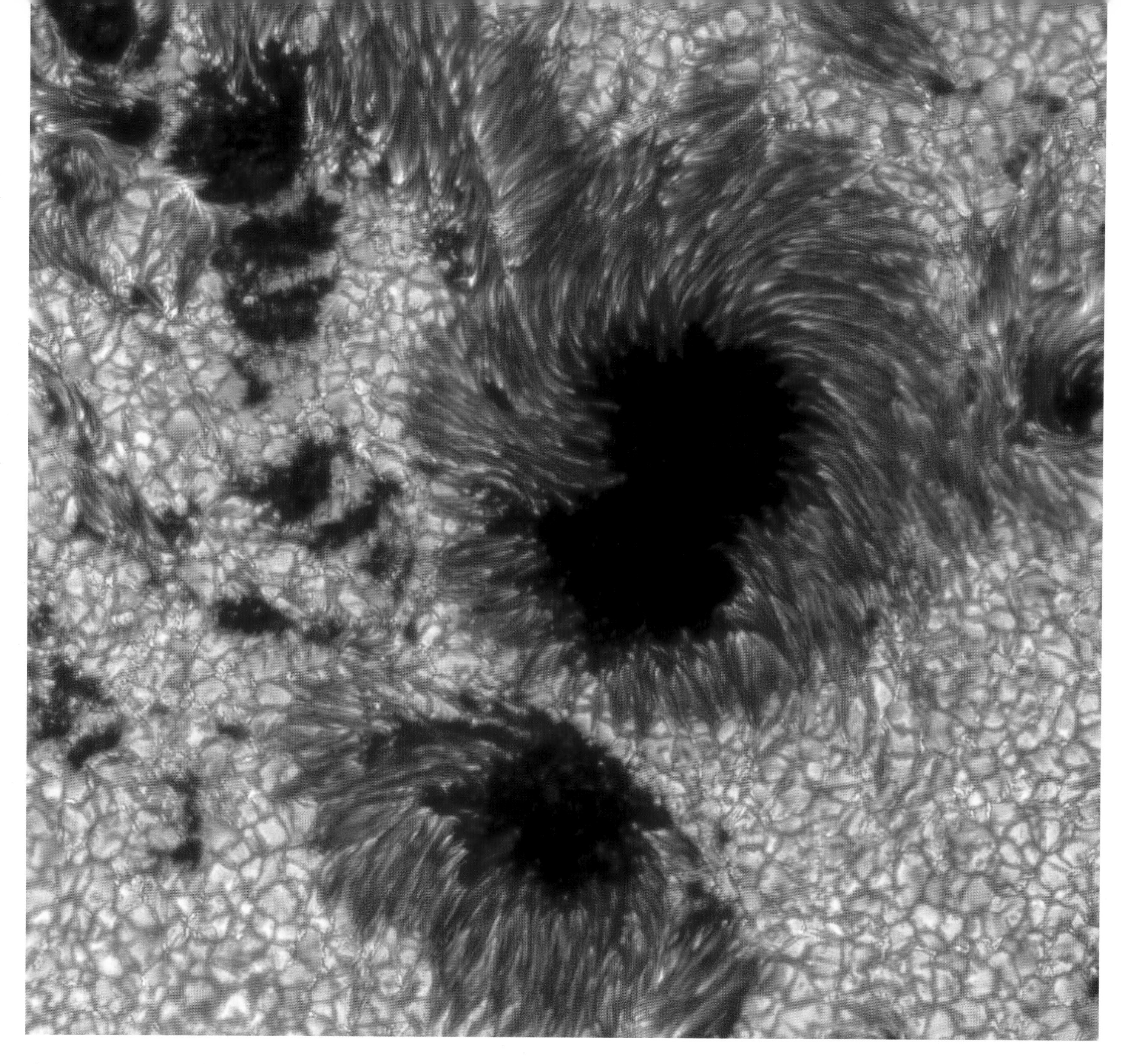

These massive stars live fast and die young. They flood the galaxy with 50,000 times the Sun's radiance for a few million years, then blast themselves into oblivion. A Deneb or a Rigel is a conspicuous beacon even when seen from 1,000 light-years away. A star the brightness of the Sun becomes lost in the stellar throng at distances greater than 50 light-years.

But regardless of their mass and projected life spans, all stars are fired by the same mechanism that makes the Sun shine. In essence, the Sun is a giant nuclear furnace. At its core, 655 million tons of hydrogen are fused into 650 million tons of helium every second, at a temperature of 15 million degrees C. The missing five million tons of matter are converted into

This extreme close-up of a sunspot group taken by a solar telescope on the Canary Islands is the most detailed image of the Sun, the nearest star, ever obtained from the Earth's surface.

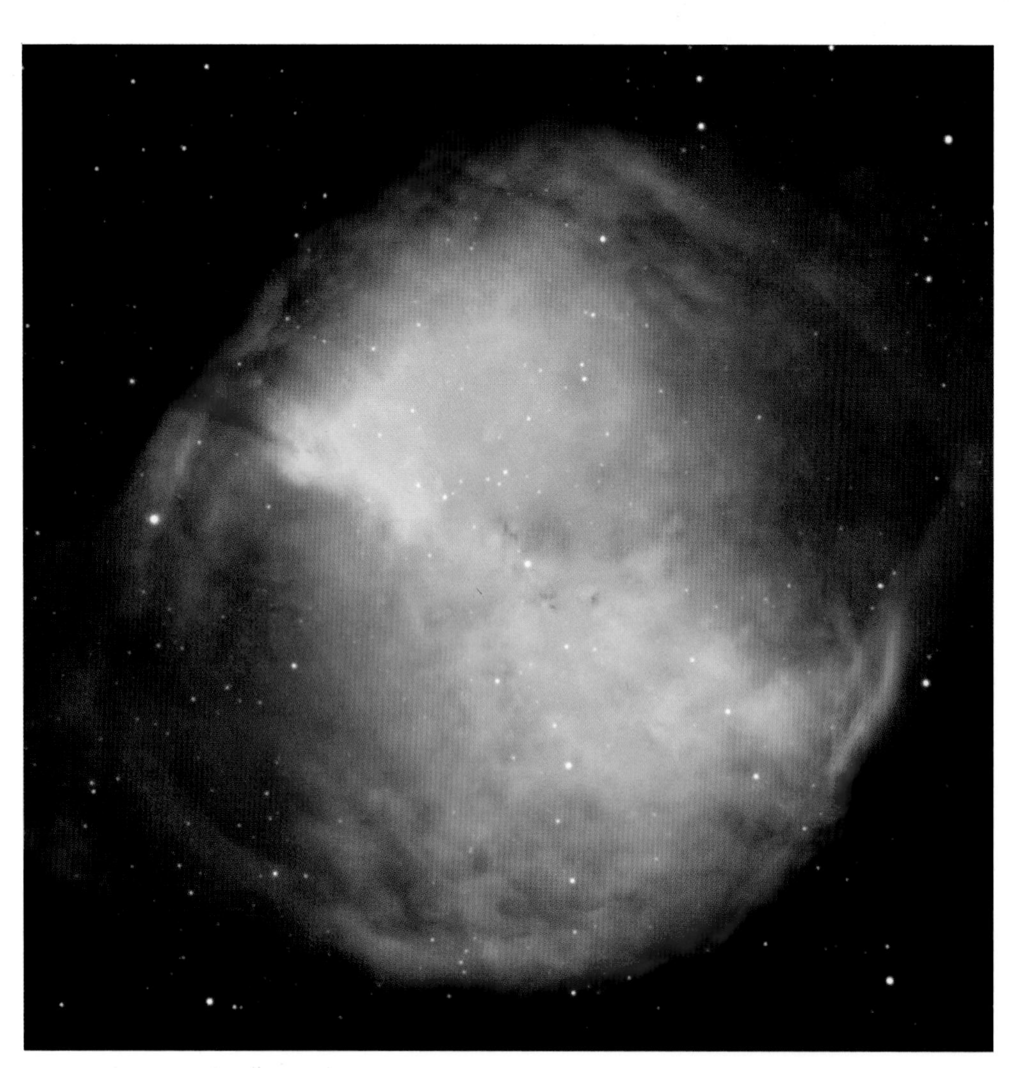

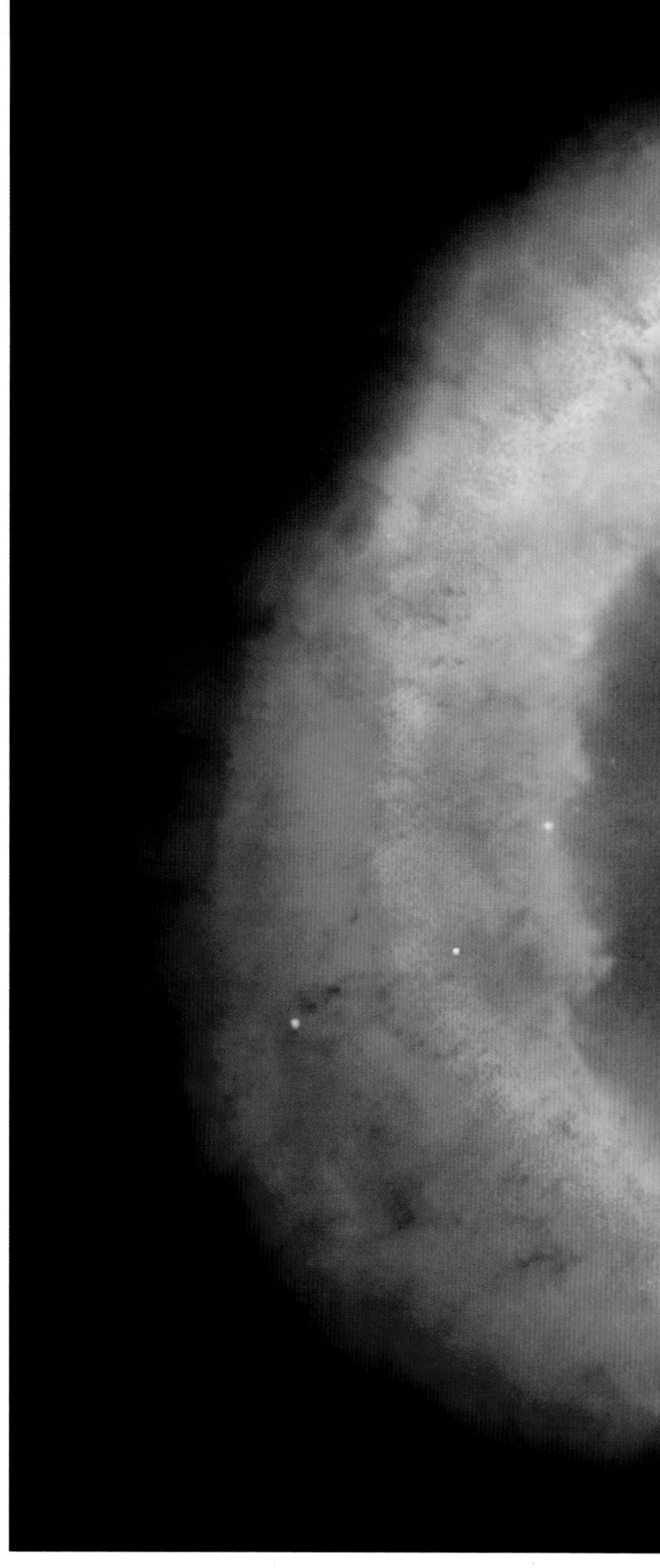

400 trillion trillion watts of energy in the process. After a tortuous trek lasting up to a million years, the core-generated energy works its way to the surface and is radiated into space, mostly as light. The Sun's steady energy output is crucial for life on Earth. An abrupt change of only a few percent in the Sun's production would vastly alter global climatic conditions. Apparently, the Sun has never had a serious power outage, or blowout. However, it will not always be so. In five to six billion years, a major disruption is inevitable.

The Sun's midlife crisis will be triggered by a depletion of hydrogen at the core. Starved for fuel to stoke its nuclear furnace, the Sun will face an energy crunch. The thermonuclear reactions will then be transferred to a shell around the core, where hydrogen will still exist. The core will contract, which will heat up the surrounding layer of burning hydrogen, accelerating the reactions and producing more energy.

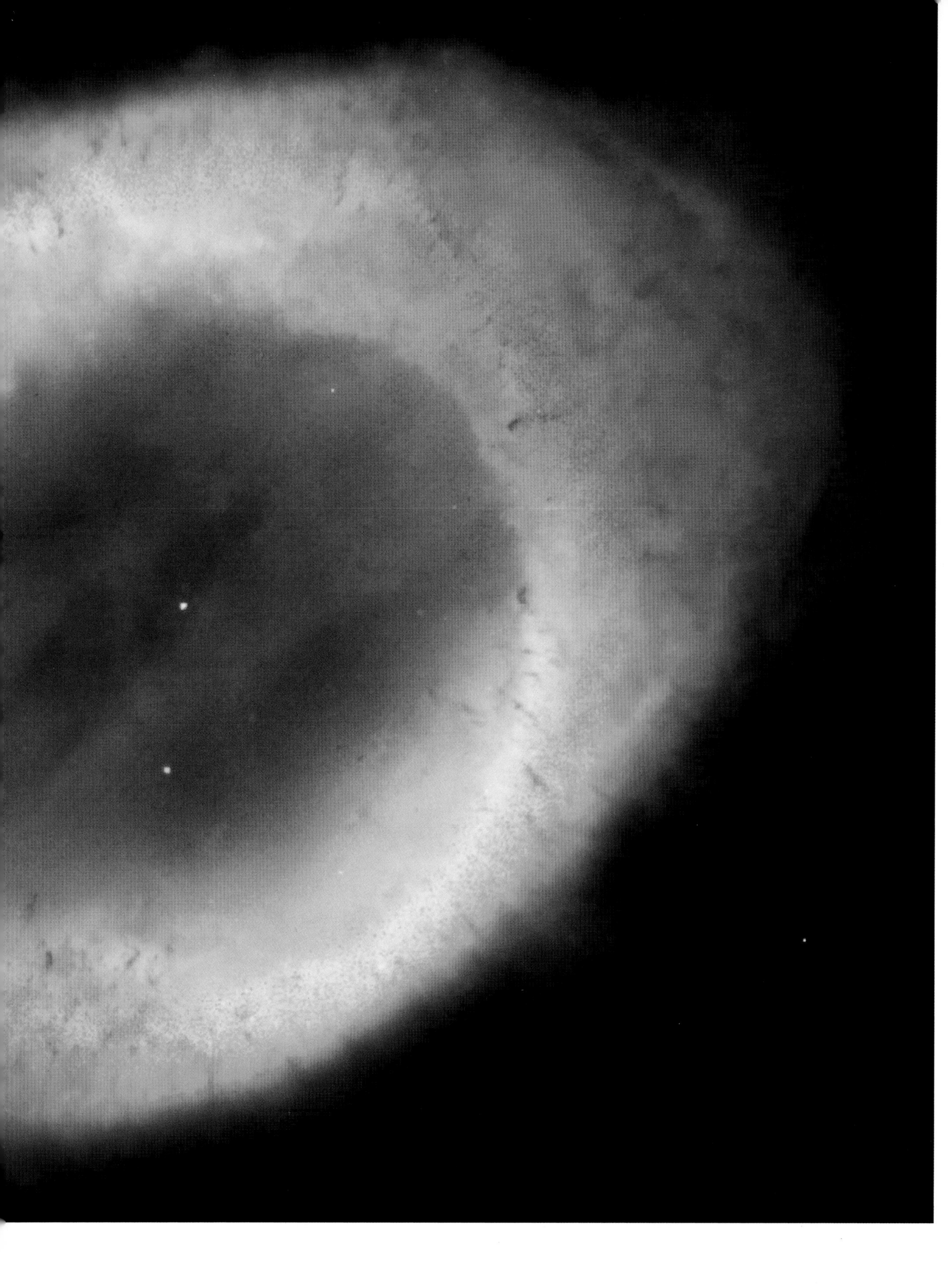

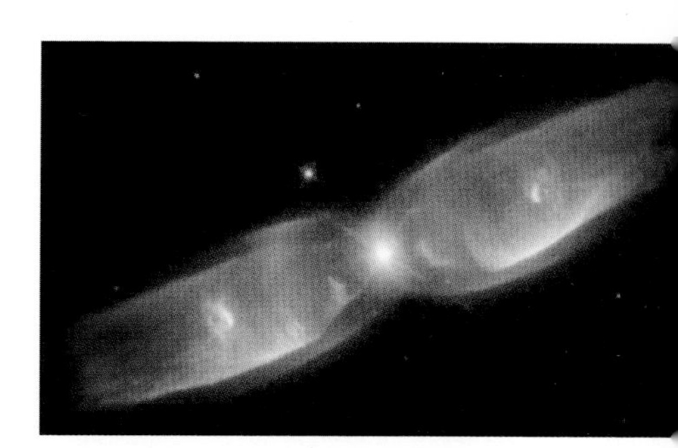

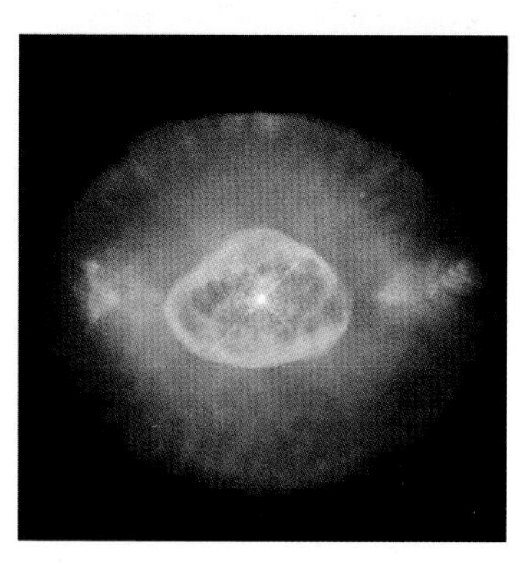

Hundreds of celestial baubles like these decorate the universe. They are the colorful last gasps of dying stars that have exhausted their cargo of thermonuclear fuel and are rapidly descending into white-dwarf senility. As white dwarfs form, they gently blow off an envelope of gas, leaving a spherical bubble illuminated like a Halloween lantern. Although they are known as planetary nebulas, they have nothing to do with planets. Astronomers in the 19th century noted that they resembled the pale greenish blue disks of the planets Uranus and Neptune through a telescope, hence the designation. At far left is the Dumbbell Nebula; at left is the famous Ring Nebula in the constellation Lyra.

Left: The Eta Carinae Nebula, the largest-known star-forming region in the Milky Way Galaxy, contains enough gas and dust to give birth to thousands of stars over the next hundred million years. Named after the star at the core of the huge cloud, the Eta Carinae Nebula is a prominent naked-eye object in southern-hemisphere skies. Above: In this photo taken in the Australian outback, the Eta Carinae Nebula is the largest pinkish patch above the trees. Streaks at the bottom of the photo were caused by transport trucks passing during the 3-minute time exposure.

The elegant spiral arms of galaxies such as M101, in Ursa Major, are sprinkled with pinkish nebulas known to astronomers as H-II regions. Hot young stars born in the past few million years illuminate the H-II regions, identifying them as celestial maternity wards.

Rather than dimming, the Sun will become brighter.

On Earth, an increase in temperature will register, perhaps only a fraction of a degree per year, but in time, the enhanced solar radiation will melt whatever ice remains in the polar icecaps and make the equatorial regions intolerable. The transition will be slow. If, for example, the process had begun at the time the Egyptians were constructing the pyramids, the changes would be detectable today by sensitive instruments but would otherwise go unnoticed. However, global climate alterations of disastrous proportions would loom several millennia in the future.

Over hundreds of centuries, escalating quantities of energy being pumped out of its core then begin to puff up the Sun like an inflating balloon. Its diameter doubles, triples, then quadruples. The energy output accelerates as the core becomes hot enough to burn helium, which fuses to carbon, forcing the outer regions even farther into surrounding space. In a few million years, the edge of the Sun reaches the inner-

Inside the Rosette Nebula star factory, deep within the Milky Way Galaxy, clouds of gas and dust churn together to create dark, floating cocoons in which stars are forming. If our galaxy could be seen from afar, as M101 is in the photo on the facing page, the entire region pictured here would appear as a tiny pink speck.

most planet. Mercury has always been a dead cinder of a world, but at this point, its rocks vaporize in the heat. Temperatures above the boiling point of water now extinguish all remaining life on Earth. The oceans vaporize to form a stifling atmospheric blanket, augmented by the noxious output from volcanic activity induced by the rising surface temperatures.

The Sun's expansion will continue for millions of years until the Earth's sky is almost filled by the deep red distended globe with a glowing heart—the engine of the doomsday scenario—only partly concealed by the bloated outer layers. At this stage, the Sun is a full-fledged red-giant star, thousands of times its original brightness. Earth will probably survive the holocaust, but only as a frazzled chunk of slag. What happens next is less certain, but theory suggests that the immense energy flow from the red-giant star will act as a stellar wind, ejecting a steady stream of dust and gas, which will eventually amount to several percent of the star's mass.

Finally, 100 million years after the crisis began, the energy-producing core of the red giant will exhaust its nuclear fuels and collapse into a dense lump —a white dwarf, a stellar corpse radiating white-hot light from the intense heat of compression. The white dwarf will have roughly three-quarters of the Sun's original mass compressed into a body the size of Earth, its atoms crushed by gravity to a state where atomic nuclei swim in a dense sea of electrons. A teaspoonful of white-dwarf material would weigh about five tons. Many other stars have already met the same fate; hundreds of white dwarfs have been discovered within a few dozen light-years of the Sun.

During its formation, the white dwarf produces a blast of stellar wind even more energetic than the wind that emerged from the red giant. This has a snowplow effect, pushing the previously expelled material into a discrete bubble about a light-year across. Like the filament in a lightbulb, the dwarf's radiation lights up the surrounding gas, creating a sphere-, doughnut- or butterfly-shaped cloud called a planetary nebula that can be seen for thousands of light-years—the star's last gasp. (Because they were described by 19th-century astronomers as being similar in appearance to Uranus and Neptune, these objects were called planetary nebulas.) After about 40,000 years, a planetary nebula disperses.

Stellar Senility Approaches

The white dwarf that will represent the Sun's old age will drift among the stars in the same path it followed around the galaxy in its more robust youth. To the cinder that was once Earth, if it still exists, the new white-dwarf Sun will be a dazzling celestial diamond shedding a twilight-level glow, but virtually

THE COLOR, SIZE AND BRIGHTNESS OF STARS

All stars are not created equal. They inherit different amounts of material from the nebulas in which they are born. Superheavyweight stars are 100 times the Sun's mass; the minimum stellar mass is 8 percent of the Sun's mass. A star's mass dictates how bright it will be, how long it will live, its temperature and its size. The diagram, right, provides this information for several dozen stars that represent a cross section of the stellar population. Although the diagram requires some explanation, it is a powerful tool and can be used to plot stars other than those indicated.

Astronomers have determined the intrinsic brightness and temperature of thousands of stars, several hundred of which are shown as black dots. The background colors represent the actual colors of the stars. The coolest stars are red, the hottest blue. The range from the extreme right of the diagram to the extreme left is 1,800 degrees C to 50,000 degrees C. Astronomers prefer to use the long-established spectral types for temperature classification. There is no significance to the letters in the letter-and-number spectral classes; it is a modern version of an old system. Each letter represents an arbitrary class. Within a spectral class, there are numeric subdivisions. In the B class, for example, a B0 star is a little hotter than a B1, which is hotter than a B2, and so on, to B9. Then come A0, A1, A2, et cetera. Spectral class is determined through examination of the spectrum of starlight. Luminosity is calculated by measuring a star's apparent brightness and ascertaining its actual brightness based on its distance from Earth. If the distance is not known, the spectrum is compared with that of similar stars whose properties are known.

Once a star's spectral type and luminosity have been established, it can be plotted on the diagram. When hundreds of stars are arrayed in this fashion, a definite clumping occurs in a gentle S-shaped curve from upper left to lower right. Astronomers call this the main sequence. Ninety percent of the stars in our galaxy are found here. The main-sequence stage is the stable hydrogen-burning period of a star's life. An individual star does not evolve up or down the main sequence but stays essentially in one place until its hydrogen fuel is exhausted and an abrupt change in internal energy production occurs. At this point, it ceases to be a main-sequence star and will move to the upper right of the diagram. In the case of our Sun, it will eventually climb to a position about midway between Aldebaran and Antares. Following the red-giant phase, a quick decline in luminosity will send it to the middle left, then down to the white-dwarf zone.

After a star contracts from its mother nebula and begins radiating energy through thermonuclear reactions, its brightness and temperature will correspond to a fixed location in the main sequence, depending on its mass. Stars the same mass as our Sun are yellow and will last about 10 billion years in the main sequence. Stars one-quarter the mass of the Sun are red and have only 1 percent of the Sun's luminosity, but they will remain on the main sequence for more than 100 billion years. Conversely, stars of greater mass are hotter and brighter and have shorter main-sequence lives.

Thus a glance at the diagram yields the stars' temperatures, sizes, masses and main-sequence lifetimes. Stars off the main sequence are another story. Their luminosities, temperatures and sizes are indicated, but their masses and the length of time they will exist in their current state vary from one individual to another. For example, the red giant Mira is roughly the same mass as the Sun, whereas the red supergiant Betelgeuse is 18 times the Sun's mass and Rigel is 30 times the Sun's mass. A further complicating factor is that the stars contain different ratios of elements (depending on the initial mix in the genesis nebula), although the primary constituents are always about three-quarters hydrogen and one-quarter helium, with a few percent of the mass taken up by all the other elements. These slight variations account for the spread of stars along the width of the main sequence.

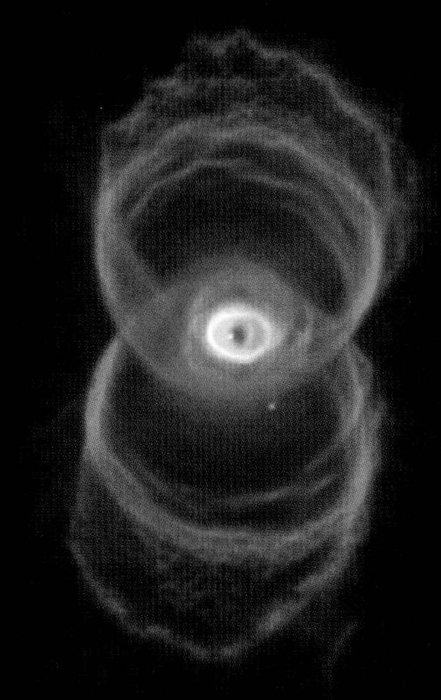

Photograph above: The Hourglass Nebula is a striking example of a solar-mass star's final gasp as it puffs off its outer layers.

Spectral Type	*Surface Temperature (degrees C)*
O5	*40,000*
B0	*28,000*
B5	*15,000*
A0	*9,500*
A5	*8,000*
F0	*7,000*
F5	*6,300*
G0	*5,700*
G5	*5,200*
K0	*4,600*
K5	*3,800*
M0	*3,200*
M5	*2,500*
M8	*2,000*

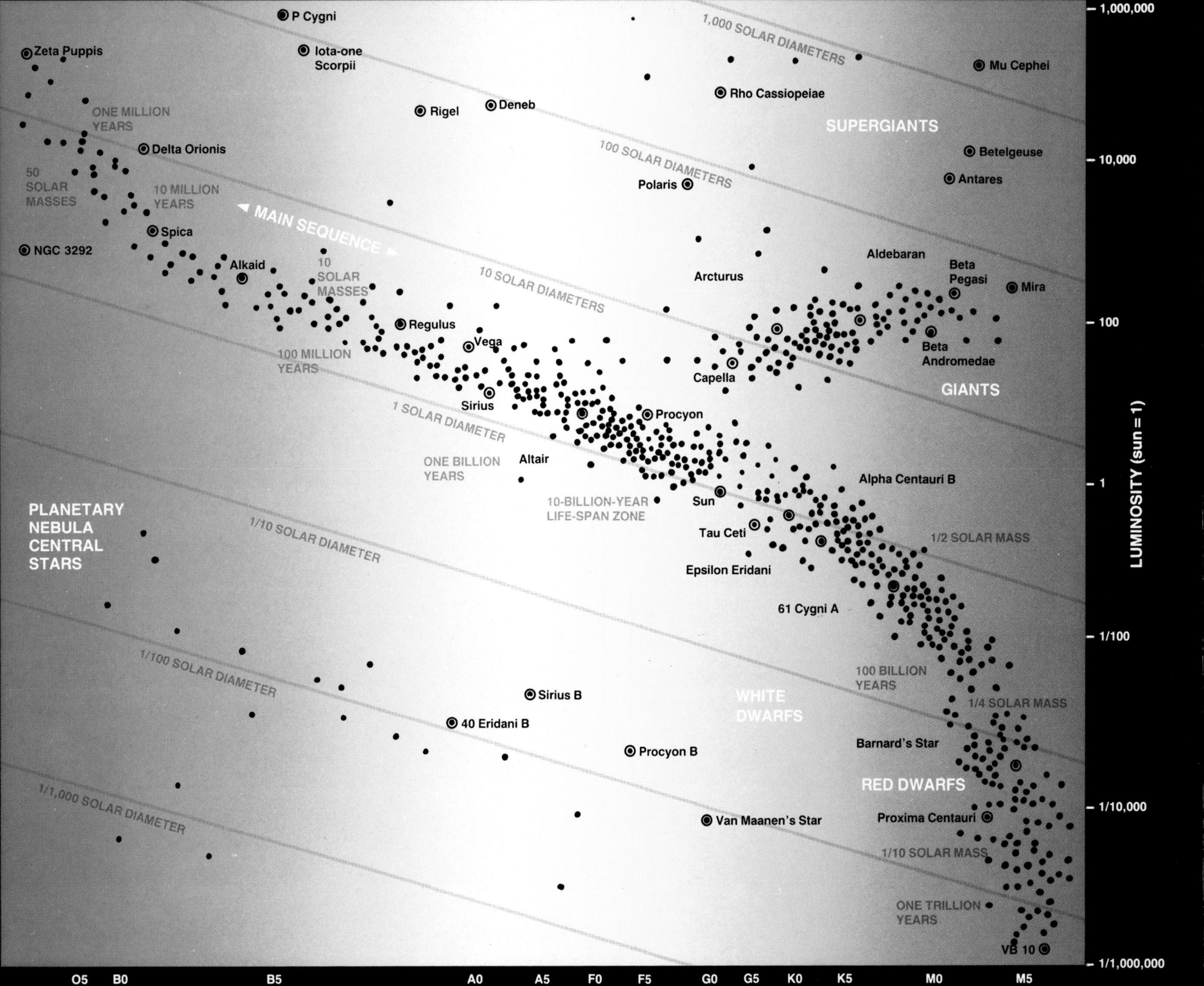

P Cygni
Zeta Puppis
Iota-one Scorpii
1,000 SOLAR DIAMETERS
Mu Cephei
Rho Cassiopeiae
Rigel
Deneb
ONE MILLION YEARS
SUPERGIANTS
Delta Orionis
100 SOLAR DIAMETERS
Betelgeuse
50 SOLAR MASSES
10 MILLION YEARS
Polaris
Antares
MAIN SEQUENCE
Spica
NGC 3292
Aldebaran
Alkaid
10 SOLAR MASSES
Arcturus
Beta Pegasi
Mira
10 SOLAR DIAMETERS
Regulus
Vega
Beta Andromedae
100 MILLION YEARS
Capella
GIANTS
Sirius
Procyon
1 SOLAR DIAMETER
ONE BILLION YEARS
Altair
Alpha Centauri B
10-BILLION-YEAR LIFE-SPAN ZONE
Sun
PLANETARY NEBULA CENTRAL STARS
Tau Ceti
1/10 SOLAR DIAMETER
1/2 SOLAR MASS
Epsilon Eridani
61 Cygni A
100 BILLION YEARS
1/100 SOLAR DIAMETER
Sirius B
WHITE DWARFS
1/4 SOLAR MASS
40 Eridani B
Barnard's Star
Procyon B
RED DWARFS
1/1,000 SOLAR DIAMETER
Van Maanen's Star
Proxima Centauri
1/10 SOLAR MASS
ONE TRILLION YEARS
VB 10
1,000,000
10,000
100
1
1/100
1/10,000
1/1,000,000
LUMINOSITY (sun = 1)
O5
B0
B5
A0
A5
F0
F5
G0
G5
K0
K5
M0
M5

The famous Horsehead Nebula in Orion is a dark silhouette against a more distant bright nebula, which is being illuminated by a hidden star. It is one of the universe's most photogenic vistas.

no heat, on a frozen wasteland. Over 30 billion years or so, the Sun will slowly cool, like a dying ember, until it no longer radiates energy.

The final corpse of the Sun will be a black dwarf about the size of Earth and 200,000 times its mass. So far, there is no observational evidence of black dwarfs. The universe is not old enough for any of the white dwarfs to have completed the slow withering death that appears to be the fate of our Sun.

Supernova: A Giant Star's Doom

Stars more massive than the Sun end their lives in spectacular fashion. It happens so abruptly, it's as if

A contorted cloud of gas and stellar debris 5,000 light-years from Earth, the Crab Nebula, top left, is a celestial tombstone marking the death of a massive star seen from Earth as a supernova in 1054 A.D. Deep within the three-light-year-wide cloud is a neutron star, the final remnant of the original star. The spinning neutron star—a pulsar—was detected from Earth in 1969. More recently, astronomers studying mysterious oscillations in the pulse rate of a pulsar 1,300 light-years from Earth—PSR 1257+12—realized that the wobbles are being caused by the gravitational tugging of two planets in orbits similar to those of Mercury and Venus in our own solar system. The planets are 3.4 and 2.8 times the Earth's mass, and they orbit the pulsar once every 67 and 98 days, respectively. The find was a complete surprise to astronomers, who had assumed that such worlds could not exist in the wake of the supernova which created the pulsar. The only reasonable theory is that the planets formed from debris left over after the supernova blast. The illustration at bottom left is an artist's rendering of the hellish conditions on the surface of the outer of the pulsar's two planets. Because of intense radiation from the pulsar, there is no possibility of life as we know it on either planet.

the galaxy were littered with cosmic time bombs. About once every half-century, a giant star somewhere in the Milky Way Galaxy annihilates itself in a supernova explosion. Most of the eruptions go unnoticed from Earth, concealed by the dust and gas in the galaxy's spiral arms. But every few centuries, a star close enough for its brilliance to remain largely undimmed by intervening clouds destroys itself in a supernova. The blast appears as a dazzling new star, sometimes bright enough to be visible in full daylight. (The last brilliant daylight supernova was in 1006.)

A supernova detonation produces a fireball a million times hotter than the surface of the Sun. Not only are such cosmic infernos spectacular, but the explosive process itself is believed to be a critically important factor in the birth of stars and galaxies and in the creation of the elements necessary for life. Not all stars become supernovas. In fact, most do not. Only those above six solar masses have the potential to go out in a blaze of glory.

Throughout its life, a star wages a battle between the force of gravity squeezing in and the outward flow of energy from its core. Massive stars, with higher core pressures, convert hydrogen to helium at a far greater

In roughly six billion years, the Sun will become a full-fledged red-giant star, bloating to 100 times its present diameter and, in the process, reducing Earth to an airless, lifeless chunk of slag. A few million years later, the Sun's outer layers will be blown off in a planetary nebula similar to those shown on pages 80-81.

rate than does the Sun and are therefore proportionately larger and brighter. In a few million years, the core of a massive star becomes clogged with helium. Gravitational compression escalates the internal temperature to burn the less efficient helium, rather than hydrogen. The star then gets bigger and hotter, and the stellar furnace is cranked up to higher and higher temperatures. The star bloats to a vast red giant, possibly wider than Jupiter's orbit. Heavier elements are created and promptly burned: carbon, nitrogen, oxygen, magnesium and, finally, iron. But iron cannot be used as stellar fuel. Iron is as inefficient as stones in a fireplace. The star becomes rotten at the core, choking on the waste products of its fuel consumption.

Now, a lethal death spiral seals the fate of the great star. Energy production ceases, and gravity takes over. The star's core implodes, while the outer layers reach temperatures of billions of degrees and erupt in the explosive fury of a supernova. Radiation floods into surrounding space and is visible millions of light-years away.

In some cases, a giant star going supernova releases as much energy as the radiation of 10 billion ordinary stars. From a distance, the giant star's death appears as a new star bursting forth where nothing was seen before. The supernova dwindles back to obscurity a few years after the eruption, but its effects are felt for centuries. The shock front from the explosion continues to surge into the galaxy like a tidal wave of energy expanding through the ocean of space. But interstellar space is not entirely empty. An exceedingly tenuous mixture of gas and dust litters a galaxy's spiral arms and gradually slows the expansion. If the supernova erupts within a few dozen light-years of a region where the gas and dust are concentrated in a discrete cloud, the shock wave wraps around the nebula, compressing it into a more compact object—like ghostly cosmic hands packing a celestial snowball. This compression is believed to be a major triggering mechanism for the creation of new stars. Moreover, the ashes of the dead star—those elements cooked up before and during the eruption—enrich the nebulas, making the element ratios in the new generation of stars different from those of the old.

All the metals on Earth were forged in the fires of massive stars. Elements lighter than iron accumulate in the lead-up to a supernova; an enriched cargo of many more elements, right up the periodic table to uranium, is created in the firestorms of supernovas. Without supernovas, elements heavier than carbon would be exceedingly rare in the universe and planets like Earth could not exist. In a universe without supernovas, it might be impossible for life to develop.

Although millions of supernovas must have illuminated our galaxy since the first stars were born, they are not common. The brightest supernova since the invention of the telescope erupted in the Earth's sky on February 23, 1987. At maximum luminosity two months later, it rivaled some medium-bright stars—easily seen but inconspicuous to anyone not looking for it. The explosion had occurred 170,000 years earlier in our nearest neighbor galaxy, the Large Magellanic Cloud, 170,000 light-years distant. A supernova seen by German astronomer Johannes Kepler in our own galaxy in 1604, about 25,000 light-years away, was as luminous as Jupiter. One observed in 1572 was only 10,000 light-years distant and shone as brightly as Venus.

There are records, mostly in Chinese chronicles, of supernovas occurring before that. One appeared in 1054 A.D. and was called a guest star by Chinese skywatchers. Today, in the exact place the Chinese records indicate, we see the Crab Nebula, a three-light-year-wide cloud of contorted star material blasted into space by a supernova. A little more than nine centuries later, in 1968, astronomers learned that the Crab Nebula is a remarkable cosmic crucible. Deep

Above: Masquerading as a hole in space, a nearby dark cloud of cosmic gas and dust several dozen light-years across blocks the light from more remote stars in our galaxy. Sooner or later, such clouds collapse to create a stellar birthplace, a celestial womb where a cluster of stars, like the one in this scene, will be born. Astronomers estimate that on average, one new star per year arises in this fashion somewhere in the galaxy. But during the past few billion years, more stars have died than have been born. In tens of billions of years, the galaxy will be far dimmer than it is today. Left: The Hubble Space Telescope reveals the bizarre remnants of Supernova 1987A, embedded in the Tarantula Nebula of the Large Magellanic Cloud. The remnants' inner pinkish ring has been expanding at one-tenth the velocity of light since the 1987 explosion. Analysis of the outer rings shows that they were emitted 20,000 years earlier by the progenitor star in its red-giant phase.

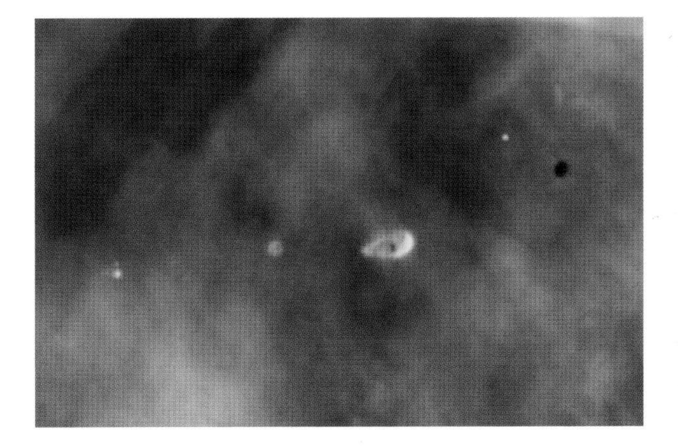

within the nebula's twisted filaments of star-stuff lies a pulsar, an object so bizarre that even though it was predicted in the 1930s, most scientists ignored the theoretical basis for its existence. The pulsar is a whirling mini-star flashing an intense, magnetically focused beam of energy like an emergency vehicle's beacon. No wider than a small city yet 500,000 times the Earth's mass, a pulsar is the superdense imploded core of a star that erupted as a supernova.

Pulsars are commonly called neutron stars because of the nature of the subatomic particles that make up these ultracompact objects. A neutron star is so alien to human concepts that it almost defies description. It is difficult to imagine how a mass of 500,000 Earths can be crammed into a sphere 15 kilometers wide. Suppose a balance scale could be built to hold a teaspoonful of neutron-star material. What would equalize the balance? Obviously, this is heavy stuff, so let's try a railroad locomotive. No, too light. Next, a million-ton oil tanker. No, that would not even budge it. No fabricated object is massive enough to tip the scales. It would take a three-kilometer-high mountain to equal the mass of this tiny amount of unearthly material.

It is easy to understand why, prior to the actual detection of a neutron star, some scientists preferred to dismiss the concept as a theoretical curiosity unrelated to the real universe. But neutron stars do exist, and their mighty magnetic fields focus energy beams that flash as pulsars when they are young and rapidly rotating. Like whirling tops, pulsars eventually spin down. Their pulsing, which gains much of its energy from twirling the magnetic field up to near lightspeed, dies out.

A deactivated spun-down neutron star would be a less formidable body to visit than one that is rapidly spinning and soaked with radiation in its pulsar phase (the Crab Nebula pulsar spins 30 times per second). Even so, there are awesome powers associated with older neutron stars. A several-billion-year-old neutron star would spin on its axis about once a minute, still a dizzying rate for something that is the size of a small city. At close range, it would resemble a white-hot ball bearing. Although in theory a spacecraft could approach such an object, a landing could never be made. The enormous surface gravity would instantly crush a spaceship and its occupants into a puddle of subatomic particles. Mountains on a neutron star would be measured in millimeters, not kilometers, even though they would contain the same amount of matter as mountains on Earth.

Following the discovery of the Crab Nebula pulsar, astronomers sought other pulsars associated with supernova remnants. The debris of past supernovas litters the sky like static puffs of smoke, but despite the detection of hundreds of pulsars, only two have

Three views of the Orion Nebula display the cosmic richness of this immense stellar maternity ward, where thousands of stars have been born over the past few million years. Facing page: A wide-angle photograph of the entire visible nebula. Radiation pressure from the brilliant stars at its core, known as the Trapezium, has pushed open the cloud like a blossoming flower. Above: A close-up of the Trapezium area. Above left: A Hubble Space Telescope extreme close-up reveals four dense knots —nebulous cocoons nurturing new stars and perhaps their attendant planets.

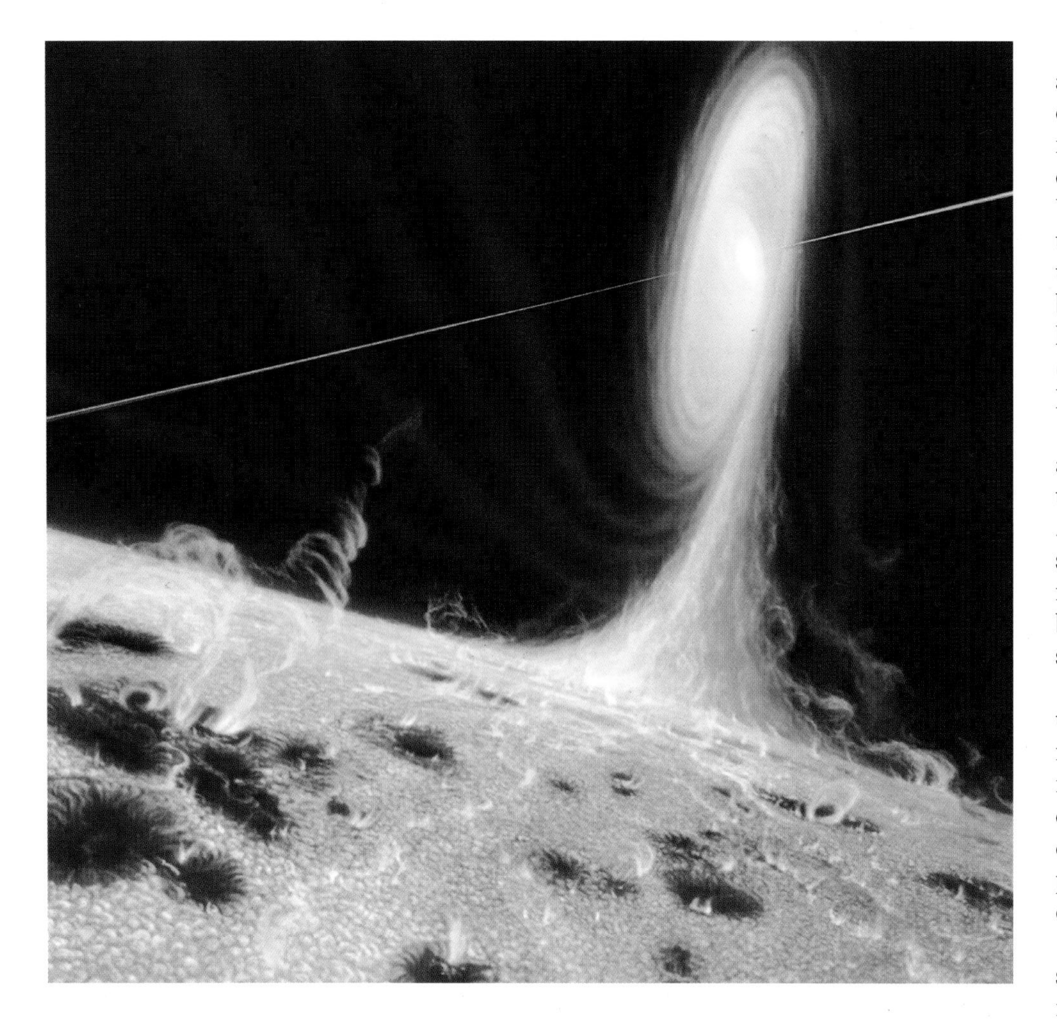

In this illustration, a black hole orbits a blue-giant star that provides generous servings of gas for the hole to devour. As the gas swirls into the black hole's accretion disk, excess material squirts out in two jets at an incredible 100,000 kilometers per second—one-third the velocity of light.

been found that are obviously embedded in matter expelled by a supernova: the Crab Nebula pulsar and one at the center of a large tattered ring of gas in the constellation Vela. The supernova that created the Vela pulsar occurred between 6,000 and 11,000 years ago, an approximation based on estimates of the supernova remnants' velocity as the gas and dust hurtle away from the site of the explosion. Astronomers calculate that the Vela pulsar is just over 1,000 light-years distant, one-quarter the distance of the Crab Nebula pulsar. At that distance, the Vela supernova must have been as bright as the first-quarter Moon, far more dazzling than Venus.

Black Holes: Gravity Whirlpools

Gravity, the master architect of the universe, has always fascinated astronomers. Orbits, masses, densities, velocities—all are controlled by gravity. John Michell, an English clergyman, amateur astronomer and geologist who became famous for his invention of the torsion balance and for the creation of seismology as a science, attempted the first predictions of the ultimate extension of gravity's power in 1784. Using Isaac Newton's formulas, Michell calculated that an object must attain one twenty-five-thousandth the speed of light to escape the Earth's gravity. To break away from the Sun's gravity, a particle would have to achieve one five-hundredth the speed of light (light's speed, 299,792 kilometers per second, was known with reasonable accuracy at that time).

Michell extended his musings one step further in a letter to Cambridge physicist Henry Cavendish. If the mass of the Sun were to be increased by a factor of 500, he suggested, the escape velocity would equal the speed of light. He concluded that "all light emitted from such a body would be made to return toward it by its own proper gravity." The more massive star, he surmised, would be invisible to distant observers.

Little could be added to Michell's concept until the 20th century. In 1916, shortly after Albert Einstein formulated the theory of general relativity, his German colleague Karl Schwarzschild realized it predicted that a star of sufficient compactness and density would be crushed by its own gravitational pull. Only the gravitational field would remain; the object itself would disappear.

For a celestial body of the Sun's mass, Schwarzschild calculated the critical diameter at about six kilometers. Once a solar-mass body collapses past that size, it has dug itself a black hole in space and disappears into it. This critical size, called the event horizon, marks the black hole's boundary. It varies in direct proportion to the mass of the object that collapsed to create the hole.

A black hole with the Earth's mass would have an event horizon the size of a golf ball. But our planet cannot become a black hole. Neither can the Sun. They are not massive enough for gravity to have the last word. Earth is as compressed as it will ever get, and the Sun will not shrink past the white-dwarf stage. But stars more than 1.4 times the Sun's mass can be compressed to incredible densities as neutron stars. Neutron stars have their limits too. A collapsing stellar remnant more than four solar masses will cascade past the neutron-star stage to the ultimate doom of a black hole. A massive star is just a black hole waiting to be born.

A black hole forms in the same type of supernova fireball that creates neutron stars. If the exploding star exceeds about 10 to 15 times the mass of the Sun (the exact figure remains uncertain), the remnant imploding core can be more than four solar masses. If it does exceed the critical four-solar-mass threshold, gravity takes ultimate control, crushing atomic particles into each other with such fury that nothing can stop the infall. The matter that creates the black hole disappears, leaving only the gravitational field. Like the Cheshire cat in *Alice in Wonderland*, all that remains is the disembodied grin of its gravity. From afar, the black hole's gravity has the same effect on objects in space as it did when the original matter existed, but closer in, the gravitational force soars, becoming so great that it prevents the escape of light.

Although astronomers suspect that supermassive black holes exist at the cores of all galaxies, only a few dozen black holes approximately the Sun's mass have been identified within the Milky Way Galaxy. Cygnus X-1, the first and, in many ways, the most convincing example, was discovered in 1971, when its intense x-ray emissions were recorded by Uhuru, the first orbiting x-ray observatory. (The "sighting" had to await orbiting instruments because the Earth's protective atmosphere screens out all celestial x-rays.) The x-ray source coincides with the position of a star known as HDE 226868, a blue-giant star about 27 times the Sun's mass at a distance of 11,000 light-years.

Spectroscopes on Earth-based telescopes revealed that the blue giant approaches then pulls away from Earth over a 5½-day cycle, indicating that it is in a 5½-day orbit around an invisible object. By analyzing this orbit, astronomers calculated that the object must be roughly 15 times the mass of the Sun. Since the maximum for a neutron star is four solar masses, there seems to be no alternative to a black hole.

The two objects in the Cygnus X-1 system are one-fifth of an astronomical unit apart (1 AU = Earth-Sun distance), and the black hole is swallowing up matter from the blue giant at the rate of 100 billion tons a day. As it plunges into the hole, the material is cooked at hundreds of millions of degrees, resulting in the x-ray emissions that first drew astronomers' attention. Matter swirling around the black hole forms a flat disk at the hole's equator (the momentum of the original star's rotation has been preserved, giving the hole an equator and poles). Accretion disks, as they are called, are thought to exist around all high-density objects—black holes, neutron stars and white dwarfs—that are pulling matter from their surroundings. When such a high-density object is coupled with a normal star in a close binary system, prodigious amounts of high-energy radiation are generated as the compact object heats, then gobbles up material from its neighbor.

A black hole several thousand times the mass of the Sun lurks near the core of our galaxy in this illustration, sweeping up interstellar gas and methodically churning it into an accretion disk. As the gas plunges into the hole (invisible at the core of the accretion disk), the excess spills into two jets that propel matter above the hole's rotational poles.

Close-Up of a Black Hole

Since their discovery, black holes have posed an intriguing question: What are they like close up? Because of the intense accretion-disk radiation, it would be lethal to venture anywhere near a system like Cygnus X-1. However, a 10-solar-mass black hole without a companion star could be approached to within a few million kilometers. Lacking the food supply from a neighboring star and the energy emis-

sions that go along with consumption, the hole would be truly black, impossible to see. But it would still have the same gravity as a star 10 times the Sun's mass. Explorers in the vicinity of this object would need sensitive instrumentation to pin down its precise location. A black hole of 10 solar masses is just 60 kilometers in diameter.

As the exploration ship approaches the black hole, it establishes an orbit around it at a safe distance, then drops instrumented probes with accurate clocks into inward-spiraling orbits. The probes at first report nothing unusual, reacting exactly as if they were in orbit about a star of similar mass. Only when the probes close in to within the Earth-Moon distance does the power of the black hole become unmistakable. From here, the hole's gravitational force simultaneously compresses and stretches any normal substance as if it were modeling clay.

At 150 kilometers from the surface of the hole (the event horizon), the clocks on the space probes appear to run slow by about 15 percent. This time warping is caused by the hole's enormous gravitational field pulling on the signals being sent back by the probes. The effect increases until, at the event horizon, any object, whether spiraling down or going straight into the hole, is accelerated to the speed of light before finally plummeting in. Once the probes cross the event horizon, their signals, even at the speed of light, cannot escape. The hole is a one-way tunnel into a substratum of space from which there is no return.

Anything that enters a black hole leaves the universe—*our* universe, at least. To our perceptions, it is gone. Nothing inside a black hole can communicate with the universe it left. No known force can break a hole's grip. A black hole is omnivorous, consuming anything. It is a one-way trap in time and space. Whatever crosses the event horizon is stretched spaghetti-thin, pulverized by gravitational tidal forces and sucked into the singularity, the black hole's heart.

Because the structure of black holes is not fully understood, there is no consensus about the ultimate fate of material consumed by these gravity whirlpools. Is it pumped out of the universe forever through "white holes" that fountain the matter back into the universe at another time and place? Are black holes the key to the wormhole space-warp drives that prop up so much of science fiction? Many astrophysicists maintain that none of these possibilities is likely. Black holes may simply be what they appear to be: the universe's ultimate abyss.

Viewed from a distance of less than one light-year, the nucleus of the Milky Way Galaxy reveals a waltz of doom orchestrated by the laws of gravity in this artist's rendering. Three powerful black holes pivot around a supermassive black hole (four million solar masses) that will ultimately devour the others, adding their mass to its own. Abundant gas in the galactic core forms ghostly tendrils as it swirls into the holes. The high-energy radiation released as the holes consume the gas has been detected on Earth, 28,000 light-years away.

GALAXIES

CHAPTER SIX

On clear, moonless evenings in late summer, a ghostly ribbon arches across the canopy of darkness. That delicate powdering on black velvet, known since antiquity as the Milky Way, is the visible portion of the gigantic pinwheel-shaped galaxy we inhabit, a system so vast that the unaided eye reveals only a tiny fraction of its bulk. What we see as a sky full of stars is our cosmic neighborhood, a small corner of the galaxy. The Sun resides on the inside edge of one of the curving spiral arms, about two-thirds of the way from the galactic center, in what the late astronomer Carl Sagan called "the galactic boondocks." Stars in our vicinity are, on average, seven light-years apart. The galaxy's total population of stars is at least 200 billion. Billions of other galaxies of comparable dimensions exist.

Galaxies are the cells of the immense cosmic body we perceive as the universe. Stars, planets, comets, gas, dust and nebulas in profusion are the galactic ingredients. Our Milky Way Galaxy is a typical spiral galaxy: an enormous disk about 90,000 light-years from edge to edge, with a bulge at the hub that is 10,000 light-years thick. In our region, 27,000 light-years out from the nucleus, the galaxy is less than 2,000 light-years thick. Two CD disks glued together make a correctly proportioned analogue of the galaxy's width-to-thickness ratio, apart from the central bulge. The glue represents the 200-light-year-thick region where nebulas and newborn stars are concentrated in the central plane of our galaxy. The Sun stays within this zone for most of its nearly circular 200-million-year orbit around the galactic nucleus, regularly oscillating slightly above or below it.

There are three basic types of galaxies: spirals, like the Milky Way; spherical swarms of stars called elliptical galaxies; and irregular galaxies, loose collections of stars with no distinct structural form. The Large and Small Magellanic Clouds, the Milky Way's satellite galaxies, are irregulars. Two nearby ellipticals are companions of the Andromeda Galaxy and are easily seen in photographs. Most elliptical galaxies are like red-dwarf stars—ubiquitous and easily overlooked. The smallest elliptical galaxies, called dwarf spheroidal systems, have only a few thousand stars. But just as red-giant stars are immensely larger than red dwarfs, giant ellipticals are the titans of the cosmos, with populations of up to 50 trillion stars. Spiral galaxies have a comparatively small size range. The largest are less than 10 times the size of the Milky Way, while the smallest are about one-fifth of our galaxy's diameter. Although spiral galaxies are undoubtedly outnumbered by ellipticals, spirals appear to be the most common, because a typical spiral is larger and brighter than an average elliptical. Small, inconspicuous irregular galaxies probably outnumber both ellipticals and spirals.

The elegant pinwheel form of spiral galaxies resembles the whirlpool configuration of a hurricane in a satellite photograph or the pattern produced by a dual-jet rotating water sprinkler. Although these models may look superficially like galaxy structures, neither offers a suitable comparison. The hurricane is an intense low-pressure area where the winds swirl in to a focus; the sprinkler jets force water out from a central point. The spiral galaxy structure, however, is thought to be the result of gravitational density waves, similar in some respects to ocean waves.

Just as ocean waves propagate across the surface

The human search for the structure of the universe is more important than finding it.

Joseph Silk

An island of more than 200 billion stars floats in the blackness of deep space 15 million light-years from Earth. This galaxy, known only by its catalog number M83, is roughly the same size and mass as our own Milky Way. Above: The eruptive galaxy M82.

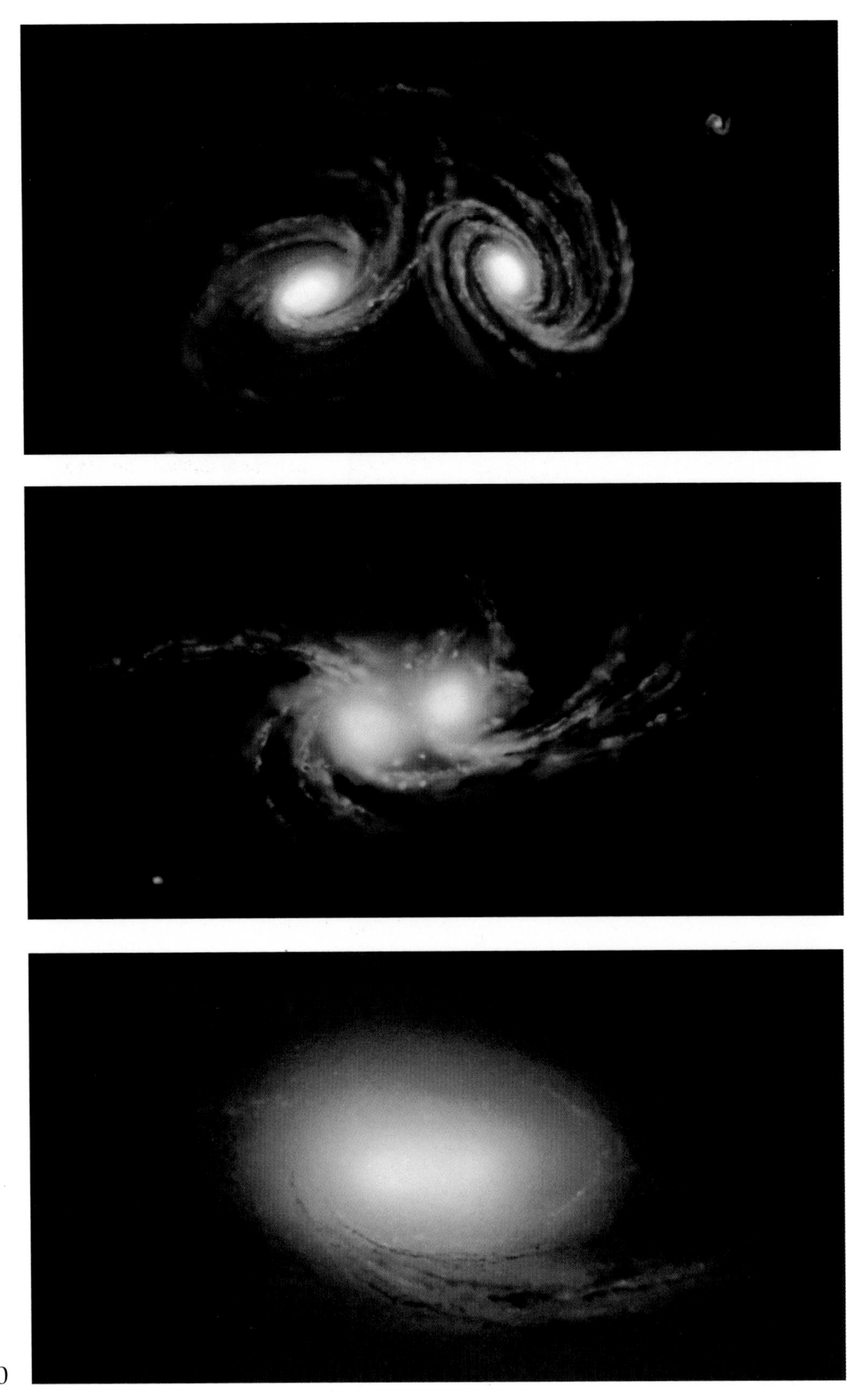

of the water but do not reflect the motion of the entire body of water, galaxy density waves move independently of the galaxy's stars but in the same direction that the stars revolve around the galactic nucleus. In the Milky Way Galaxy, the density waves advance at approximately half the speed of the stars moving around the nucleus. The stars and nebulas in a density wave today will be between waves millions of years from now and will enter a new wave millions of years after that. The visible spiral arms mark regions where density waves produce a modest compression of the dust and gas.

To visualize how a density wave affects the region of the galaxy it occupies, imagine three lanes of heavy traffic on a freeway encountering a slow-moving truck. The cars bunch up, eventually pass the truck and resume normal speed. From a helicopter above, the blob of cars behind the truck would be an automobile density wave moving along the freeway more slowly than the normal flow of traffic. A galaxy's density wave seems to be a self-propagating zone where stars and nebulas are moderately bunched at one particular time. Since stars are far apart to begin with, the density wave has a negligible effect on them. It is the nebulas—gas and dust clouds tens of light-years wide—that "feel" the density wave.

The compressing of nebulas can set in motion a chain reaction of gravitational collapse in the densest parts of the clouds, triggering bursts of star formation. The newborn stars illuminate the nebulas around them, as is the case with the Orion Nebula. These glowing clouds of gas and dust, called emission nebulas, appear pink in photographs of the Milky Way or of the spiral arms of other galaxies.

The biggest and brightest young stars that form in the density-wave turbulence are blue giants thousands of times more radiant than the Sun but with life spans of just a few million years. These stars give galaxy spiral arms their predominantly blue color. Before the density wave moves on, the largest blue giants (the "live fast and die young" stars) expire in supernova explosions whose shock waves ripple through surrounding space and compress other nebulas. The result is a chain reaction of star birth up and down the density wave. Density waves, though themselves invisible, create the visible spiral arms—distinct regions knotted with young blue stars and pinkish emission nebulas that, together, make the spiral arms far more luminous than the regions between the arms.

As the density wave moves on, the most massive and brightest stars burn out, the nebulas cease to be

Facing page: The collision and merger of two spiral galaxies—depicted at billion-year intervals—can be a messy affair. The final result of such cosmic smashups is usually a single elliptical galaxy. On this page, the Hubble Space Telescope has caught one galaxy plunging through the middle of a neighbor (top), while a second pair of galaxies in collision (bottom) has sideswiped each other, flinging tails of billions of stars into the abyss. Appropriately, these two galaxies have been dubbed The Mice.

Facing page: The Large Magellanic Cloud, our nearest galactic neighbor, is littered with stellar nurseries. The largest, the Tarantula Nebula, at upper left, is 10 times wider than the Orion Nebula. The Large Magellanic Cloud may have originally been a barred spiral, but the Milky Way Galaxy's gravitational muscle has warped it into ragged submission. This page, top: Both the Large and Small Magellanic Clouds are visible to the left of the observatory in this image, taken at Cerro Tololo in Chile. A section of the Milky Way is visible at right. The Large Magellanic Cloud is about 5 percent of the mass of the Milky Way Galaxy, while the Small Magellanic Cloud is less than 1 percent of our galaxy's mass. The Small Magellanic Cloud, bottom, is another irregular galaxy companion to the Milky Way. To its right is the globular cluster 47 Tucanae.

illuminated and the region darkens until the next density wave comes along. The interarm area is still star-rich but appears almost empty, because the hurly-burly of star formation is largely confined to the density wave. Our Sun is on the inside edge of the Cygnus-Orion Arm of the Milky Way Galaxy. Since the Sun's velocity is twice that of the Orion Arm density wave, we should be passing out of it within the next 10 million years.

A second theory to explain spiral arms suggests that the collision of large nebulas in the galaxy starts a chain reaction of star formation, supernovas and more star formation. The cycle continues for a few hundred million years, during which the rotation of the galaxy twists the star-forming region into a long spiral shape. Both theories have their adherents, but there is still much to learn about how the immense star cities have become stabilized in their dazzling pinwheel structures.

Galactic Contrasts

Spiral galaxies are undoubtedly among nature's most elegant creations. Their graceful curving arms, traced by millions of blue-giant stars and puffs of pink nebulosity, contrast with the bulging nucleus where 100 billion stars are nestled.

Compared with spirals, elliptical galaxies are rather dumpy. Ellipticals are star piles, vast spherical or football-shaped blobs of yellow and red stars. Although not ugly, ellipticals certainly lack flair.

Why did different types of galaxies form after the Big Bang? Why are they not all spirals?

Until the early 1980s, astronomers assumed that as the universe's primordial gas clouds collapsed into galaxies one to two billion years after the Big Bang, a galaxy's shape was determined by the manner in which the clouds collapsed. Fast-spinning clouds produced spirals; slower-spinning material collected as ellipticals. The dregs became the irregulars. But this idea is no longer favored. There is growing evidence that the genesis galaxies were almost entirely spirals.

In the new view, ellipticals and irregulars are the wreckage of spirals that collided within a few billion years of forming. At their formation, say, one billion years after the Big Bang (13.7 billion years ago), the universe was a relatively crowded place. Less than one ten-thousandth its present volume, the universe had the same amount of matter then as now. That would be like taking a large two-story house and reducing it to the size of a doghouse suitable for a dachshund. The newborn galaxies were almost rubbing shoulders, and collisions would have been far more common than they are today. This is, in fact, what the deepest Hubble images reveal. Many galaxies at the limits of observation, which we are seeing as they were billions of years ago, have contorted shapes—just what would be expected.

A galaxy collision would last for hundreds of millions of years. Once two spirals actually touch, they are likely doomed to loop ever closer to each other before finally merging. Yet because of the distance between individual stars, stellar collisions would be exceedingly rare. If the Milky Way Galaxy were to collide with the Andromeda Galaxy, the chances are slim that a star from Andromeda would even come close enough to knock one of the solar system's planets out of orbit. From the point of view of Earthlings, a collision with Andromeda would pose less of a threat to our planet than would a hit by an errant asteroid or comet from within our own solar system.

On the other hand, galactic mergers create quite a fuss on a large scale, because collisions among the abundant gas clouds in the young galaxies would trigger star formation. Friction from the cloud collisions, plus dynamical friction from the gravitational interaction of the two great star cities, would rip apart the spiral arms and cascade the remains into a chaotic mass of stars—an elliptical galaxy.

Even a close miss would twist off much of a spiral galaxy's twirling arms, although the nuclear region would remain essentially intact. In such a gravitational tug-of-war between two continents of stars, the outrider stars would receive two sets of signals that would scramble their orbits. As a result, many billions of stars would be flung into intergalactic trajectories and lost from both galaxies forever.

Galaxy collisions are relatively rare today, although Centaurus A (shown on page 113) is the debris from two nearby galaxies that collided two billion years ago. At least one of the original pair was a spiral. Elliptical galaxies, it seems, are not born—they are merely heaps of stars from dismembered spirals.

Journey to the Galaxy's Center

Out in the Cygnus-Orion Arm of the Milky Way Galaxy, where the Sun resides, there is one star for every 400 cubic light-years. The next spiral arm inward, 6,000 light-years from us, is the Sagittarius Arm, which is more generously endowed with stars and nebulas than is our arm of the galaxy. From deep

Facing page: At a distance of 25 million light-years, The Antennae is a nearby example of two colliding galaxies that have nearly merged into one. Once spirals, the galaxies have intertwined to the extent that their individual structures are masked by a vast arm of intense star and cluster formation caused by the chaotic collision of nebulas. The two bright yellow regions are the cores of the original galaxies. Above: Although this pair of galaxies appears to be involved in an intergalactic smashup, one is slightly in front of the other. Nevertheless, a full-blown collision is inevitable as their mutual gravity pulls them into a deadly embrace. Yes, the universe is expanding, but not over relatively small distances like this.

DISTANCES IN THE COSMOS

More than a dozen methods are used to measure distances in the universe. The main ones are given in the table. Among the various techniques, Cepheid variable stars are the key distance calibrators. The intrinsic brightness of each Cepheid is directly related to its rhythmic oscillations in luminosity. For instance, a Cepheid with a long period of oscillation has a high intrinsic brightness, like a 100-watt lightbulb, whereas a short-period Cepheid would be a 40-watt bulb, and so on. Comparing the real luminosity, or wattage, with the apparent brightness reveals the distance. Fortunately, Cepheids are typically more than 1,000 times brighter than the Sun and can be seen across vast distances. For instance, the Hubble Space Telescope portrait of the galaxy M100 on the facing page has revealed dozens of Cepheids. One of them is shown in the three inset images, taken over a 10-day period. If you glance from one image to another, you can see that the Cepheid changes brightness. By measuring the periods of variation and brightness of more than two dozen Cepheids, astronomers have determined that M100 is 51 million light-years from Earth.

Distance	*Objects*	*Method*	*Technique or Assumptions*	*Accuracy*
up to 50 AU	*planets, asteroids and comets*	*radar*	*powerful radar aimed from Earth is bounced off object; round-trip time gives distance*	*can be accurate to a few meters*
4 to 200 light-years	*nearby stars*	*heliocentric parallax*	*photos of star taken when Earth is on opposite sides of its orbit; shift of star compared with background stars yields distance*	*accurate to less than 1 light-year for stars within 30 light-years; loses precision rapidly after 200 light-years*
100 to 200,000 light-years	*stars in our galaxy and Magellanic Clouds*	*main-sequence fitting*	*a star's spectrum often tells where it fits on the main sequence and thus its true luminosity (see page 87)*	*considered accurate within plus or minus 20%, though many giant stars are tricky to assess*
to 70 million light-years	*nearby galaxies*	*Cepheid variable stars*	*true luminosity of Cepheids is directly related to the star's period of variability*	*regarded as the most important "standard candle" in the universe*
to 200 million light-years	*nearby galaxies*	*blue supergiant stars*	*the very brightest blue supergiants in a galaxy are about the same luminosity as those in our galaxy*	*not as accurate as Cepheids, but blue supergiants are brighter*
to 500 million light-years	*nearby galaxies*	*globular clusters*	*the brightest globular clusters in a galaxy are about the same for all galaxies*	*not as accurate as above methods, but useful as a cross-check*
to at least 1 billion light-years	*nearby galaxies*	*type Ia supernovas*	*type Ia supernovas are all believed to have nearly the same maximum luminosity*	*because type Ia supernovas are brighter than anything else in a galaxy, this method reaches farther out*
to several billion light-years	*galaxies*	*galaxy brightness*	*several techniques are used to estimate a galaxy's true luminosity*	*distance is essentially an educated guess*
to several billion light-years	*galaxy clusters*	*brightest galaxies*	*the most massive galaxies may be similar from cluster to cluster*	*useful when galaxies are so faint and remote that no other method works*
100 million to 10 billion or more light-years	*remote galaxies*	*redshift*	*because of the universe's overall expansion, the faster a galaxy is receding, the farther away it is*	*overall, the most powerful tool in the astronomers' arsenal*

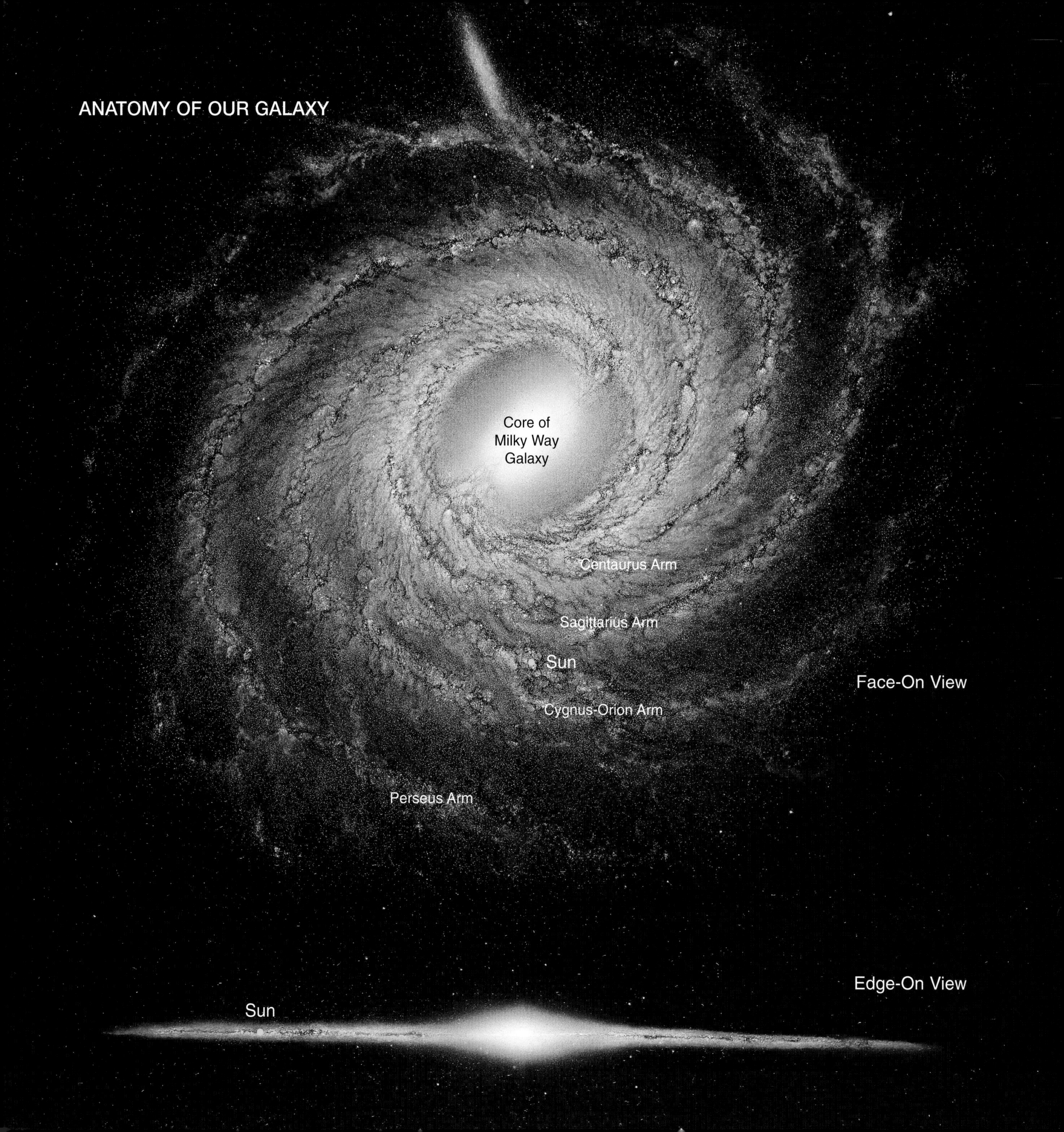

ANATOMY OF OUR GALAXY

Core of
Milky Way
Galaxy

Centaurus Arm

Sagittarius Arm

Sun

Face-On View

Cygnus-Orion Arm

Perseus Arm

Edge-On View

Sun

within the Sagittarius Arm, the galactic landscape would appear noticeably brighter than the sky in our region of the galaxy. The Centaurus Arm is next, about 8,000 light-years farther in toward the galaxy's nucleus. Emerging from the Centaurus Arm, a galactic traveler would see a wall of stars ahead—the hub of the galaxy, the galactic central bulge. From this distance (less than 10,000 light-years), the core region has a rich golden glow, the combined light of billions of stars far more closely packed than anywhere in the spiral arms. Very few blue giants are seen. Instead, the brightest stars are red giants like Betelgeuse and Antares. The vast majority of stars are yellow or orange. Unlike the most luminous parts of the spiral arm, the nucleus does not derive its brilliance from short-lived blue giants but, rather, by brute force from throngs of lesser suns whose combined light creates a dazzling scene, like hundreds of bursting skyrockets superimposed and frozen in time.

Two thousand light-years from the galaxy's center, the nucleus is still dimmed by clouds of gas and dust that mingle with the stars. At 150 light-years from the core, about twice the distance from the Sun to the majority of the stars in the Big Dipper, the galaxy's nucleus can just be glimpsed. It is an intense concentration of light, almost as radiant as the Sun appears in the Earth's sky, surrounded by countless stars, mostly yellow but highlighted by red giants.

Thirty light-years away. The nuclear region now exceeds the brilliance of full sunlight, but the heart of the galaxy is concealed from direct view by gas and dust brilliantly illuminated by 60 million stars within a 30-light-year radius.

Even at a distance of three light-years, the maw of the galaxy, veiled by gas and dust, is still not visible. In a 400-cubic-light-year zone at the galaxy's nucleus —a volume in which only a single star would be found in the spiral arms—there are more than two million stars along with liberal amounts of churning gas and dust. There is no night here. The sky contains

A comparison of the Milky Way Galaxy viewed edge-on from the inside (above, left) and a distant spiral galaxy seen edge-on from the outside (above, right). Note the similar shapes of the dust rifts in both. Facing page: Structural details of the Milky Way Galaxy are depicted in this illustration.

The center of our galaxy, right, is shielded from human vision by clouds of gas and dust. However, this infrared spacecraft view shows previously unseen details. Illustration facing page: This six-step zoom into the heart of a quasar begins at upper left within the confines of a galaxy cluster dominated by a giant elliptical galaxy. Jets emerging from the galaxy are seen in most quasars but are shown here in more detail. The frame below reveals some of the thousands of globular clusters surrounding the giant elliptical. Unseen is a halo of gas that is trapping a spiral galaxy which has ventured too near the massive elliptical. Sputters punctuating the jets indicate periods of enhanced activity at the core prior to their ejection. In the next increment, the core of the huge galaxy is exposed. Here, five black holes rip stars into light-year-long tendrils of gas that are ultimately funneled into the holes. The dominant black hole has the mass of several billion Suns. Closer to the galactic core, the fourth frame, upper right, shows a monstrous gravity whirlpool 100 times the diameter of the solar system. In the penultimate scene, we see the source of the quasar's jets, and the final view takes us to within a few solar-system diameters of the quasar engine: the powerhouse black hole, largely concealed, in the doughnut hole at center. The doughnut itself is the hole's accretion disk.

thousands of stars as bright as the planet Venus and dozens as luminous as the full Moon. The average distance between stars is just 20 times the diameter of Pluto's orbit.

At less than one-tenth of a light-year from the core, the heart of the Milky Way Galaxy is finally revealed: a monstrous vortex of gas, dust and starstuff that is swirling to its final resting place at the bottom of a black hole. Attending to the leftovers are smaller gravity whirlpools, all in a death-spiral dance around each other. It is a scene humans may never see. The entire region is an inferno of lethal radiation.

The guardian of the galaxy's central fortress is a four-million-solar-mass black hole. It is about 15 times the diameter of the Sun, and its accretion disk extends out to a distance equivalent to the diameter of Jupiter's orbit. Such a massive hole dwarfs the 60-kilometer-wide black hole in the Cygnus X-1 system. According to theory, material whirling toward a giant black hole would not plunge in immediately. Rather, it would be partially repelled and heated by intense radiation from the zone around the hole. Matter here is so compressed and is accelerated to such enormous velocities by the black hole's gravity that it emits vast amounts of high-energy radiation, usually in the form of gamma rays.

The first hint that the galactic core may consist of one or more giant black holes came in 1977, when high-altitude balloons fitted with gamma-ray detectors recorded intense gamma radiation coming from the direction of the galaxy's nucleus—exactly the type of radiation that theorists say could be generated only at the superheated inner fringes of an accretion disk spinning around a gigantic black hole. The accretion-disk model also fits with findings of radio astronomers, who mapped radiation from the galactic nucleus (radio waves can penetrate the gas and dust that block visual light from the nucleus).

Occasionally, a large amount of material, perhaps an entire star, plunges toward the hole, causing a burst of radiation, like throwing gasoline on a fire. The more distant material falling toward the hole is then blasted out into the galaxy. Indeed, radio astronomers have detected doughnuts of material ejected from the galaxy's nucleus like smoke rings, which have been attributed to periodic eruptions. Every few thousand years, the galaxy probably hiccups in this way.

Quasars: Light Fantastic

When quasars were discovered in 1963, they instantly became astronomy's number-one enigma. Measurements of their spectral redshifts, caused by the expansion of the universe, indicated that quasars were at enormous distances from Earth, close to the fringes of the known universe. Yet a quasar's radiation is emitted from a zone barely larger than 10 times the diameter of the Earth's orbit around the Sun. From such a relatively minuscule area, the quasar pumps out energy equivalent to that radiated by trillions of stars like the Sun.

There were no theories at the time to explain how so much energy could be coming from such a compact source. Astronomers wondered whether quasars

= 1,000,000 light-years
= 1 light-month
= 100,000 light-years
= 1 light-week
= 1 light-year
= diameter of solar system

A trio of galaxies apparently embedded in stars floats in the blackness. But this is an illusion. The stars are in our own Milky Way Galaxy, while the galactic threesome is a million times farther away. It's as if we are looking at distant trees (the galaxies) through a window speckled with raindrops (the stars).

might be less powerful than they appeared. Perhaps they were closer than first thought, somehow masquerading as remote beacons. In the past few years, however, the evidence has become overwhelming that not only are quasars out among the distant galaxies, they are *within* galaxies. More specifically, they are the nuclei of galaxies in a particularly violent stage of evolution. Due to their exceptional radiance, quasars can be seen at enormous distances, whereas the galaxies they inhabit are, by comparison, almost invisible.

Examination of the dim halos of light around most quasars has shown that the quasar "fuzz," as it is called, is the combined light of billions of stars similar to those found in galaxies such as the Milky Way. Astronomers conclude that quasars are rare superbrilliant cores of galaxies. Further proof comes from a few images that reveal hints of spiral arms emerging from the nuclei of some quasars.

Perhaps the most persuasive piece of evidence linking galaxies and quasars was a supernova seen in 1984 amid the luminous fuzz surrounding quasar QSO 159+730, about 1.5 billion light-years away. This was the first time a supernova had been found in a galaxy in which a quasar resides, although supernovas have been observed in normal galaxies hundreds of times during the past century. Supernovas can be used as distance gauges. Like streetlights, supernovas have approximately the same maximum luminosity. The dimmer they appear, therefore, the farther away they are. Judging from its brightness, the supernova seen in the quasar fuzz of QSO 159+730 is one to two billion light-years away, which corresponds to the distance determined from the redshift. Astronomers at last had a completely independent method by which to confirm a quasar's distance.

But identifying where quasars are does not necessarily establish what they are. The compacted focus of the quasar's titanic energy release cannot be explained by ordinary stellar radiation. No matter how powerful or massive the stars or how densely they are concentrated in the galactic core, they simply do not have the brute output of a quasar. Only one known object does: a giant black hole.

Giant Black Holes

A supermassive black hole has the potential to generate radiation 100 times more efficiently than the thermonuclear fires at the core of a star like the Sun. A black hole's power plant is fueled by anything that is sucked into the vortex. The enormous gravitational whirlpool twirls infalling matter into an accretion disk and accelerates it to nearly the speed of light. The process generates prodigious quantities of energy that burst into space as intense gamma and x-ray radiation just before the doomed matter plunges into the hole. The key to a quasar's fountain of radiation seems to be an abundant supply of matter for its black hole.

Stars and gas clouds swarm at the center of all galaxies, but to yield the kind of output emerging from the nucleus of a quasar, the equivalent of several hundred thousand Earths—or one whole star—in dust and gas would have to be consumed annually.

Above: Peering deep into the heart of galaxy NGC7052, the Hubble Space Telescope captured a 300-million-solar-mass black hole in action. The disk surrounding the hole is 3,700 light-years in diameter and contains three million solar masses of gas, dust and debris. The black hole itself is blocked from view by hot gas swirling into it, but velocity measurements of the material diving in leave no doubt that such forces could be generated only by a black hole. Left: At a distance of 22,000 light-years, Centaurus A may be the nearest colliding galaxy to the Milky Way. A billion years in the making, the merged galaxies look like an elliptical with a spiral galaxy wrapped around it—and that's probably just what it is.

A quasar black hole is feasting on an apparent banquet of infalling material not available in ordinary galaxies. However, it would not swallow whole stars. A doomed star would never make it to the black hole's event horizon in one piece. It would become entangled in the vortex of the accretion disk and shredded, or "spaghettified," as one astronomer put it. The huge amount of matter swirling into the quasar black hole (a billion trillion tons each second) would form a blazing doughnutlike whirlpool around the hole that would churn out more energy than the radiation from trillions of stars. The radiation blast wave itself would likely boil any stars that ventured near the hole. The resulting shreds of stars and tendrils of gas swirling into the hole would be incandescent, creating a terrifying scene of raw power unmatched in the universe.

A black hole with an abundant food supply is a messy eater. It grabs more than it can swallow. The surplus material appears to be ejected in two jets perpendicular to the accretion disk, presumably over the hole's rotational poles. Why the hole fails to suck in everything that approaches it is unclear, but the dual-jet phenomenon is seen in many quasars and in galaxies with active nuclei. In some cases, huge clouds of matter have been hurled millions of light-years from the black hole—far from the hole's gravitational grip—more than the distance from Earth to the Andromeda Galaxy.

It now seems likely that almost every galaxy has a central black hole created during galactic birth. An initial intense period of star formation and massive-star supernovas at the crowded galactic nucleus would have left many black holes in the 10-solar-mass range.

At the opposite end of the scale from the huge elliptical and spiral galaxies that dominate the visible universe at large are billions of dwarf irregular galaxies, like this one, floating almost unseen. Containing as few as several hundred thousand stars, the dwarf galaxies make up in numbers what they lack in size and are the most abundant class of galaxy in the universe.

These black holes would have grown by colliding and merging with each other and by feeding on the nebulas that were plentiful then. The larger of two colliding black holes would emerge with the added mass of the consumed hole. Eventually, one dominant hole would define the galactic nucleus. But a typical spiral galaxy's nuclear zone—as evidenced by our galaxy—does not create a billion-solar-mass quasar-level black hole. Something else is required.

Only a tiny fraction of galaxies contain quasars. The nearest one is more than a billion light-years away. Obviously, a combination of rare circumstances is needed to provide enough material for a monster galactic black hole to pump out quasar-level radiation. Galactic collisions may be the only phenomenon violent enough to do it. Clouds of dust and gas in the interacting galaxies would collide, lose momentum and plunge into the galactic nuclei during such an encounter. Consuming matter at the rate of up to a million Earth masses per year would soon inflate the galaxies' primordial black holes to millions and possibly billions of times the Sun's mass. Eventually, the two black holes at the cores of the colliding galaxies would merge.

A one-billion-solar-mass black hole would be about the diameter of Pluto's orbit, corresponding to the apparently small celestial nozzle from which the quasar energy is pumped. The larger the black hole, the more energy it releases, provided it has copious amounts of fuel.

The engine for quasar fireworks requires a hole of at least several hundred million solar masses that functions at full throttle in bursts lasting about a million years. At its peak activity, the black hole creates such an enormous energy flow outward that gas and dust clouds beyond the accretion disk are blown back, reducing the available fuel supply. The action then subsides for a few million years, the displaced material gradually returns, and the cycle repeats until the hole depletes the galactic core zone to the extent that maximum consumption is never resumed. It is doubtful whether a galaxy can remain in a quasar phase for more than two billion years. Thus many galaxies probably harbor giant central black holes that may have the potential to be quasars but are currently inactive.

The most remote quasar known is about 12 billion light-years from Earth. Far more quasars are seen at great distances than random distribution would dictate. Astronomers conclude that quasars were abundant during the first few billion years of the universe's existence, which is what we would expect if quasars are produced as a result of galactic collisions, because such collisions were more common then.

Emperor Galaxies

Dominating the central regions of the largest clusters of galaxies are obese giant elliptical galaxies ranging up to, and perhaps even exceeding, 50 trillion times the mass of the Sun. These are the masters of the universe and the bullies of the galaxy families. More massive than any other discrete stellar structure, they are the gravitational hubs of galaxy clusters.

Illustration left: Fountaining a geyser of plasma and gas, a five-billion-solar-mass black hole at the nucleus of a giant elliptical galaxy is seen from the surface of a hypothetical planet. Stars near the hole are turned into cometlike bodies as they plummet to their doom, eventually merging with the black hole's huge accretion disk. The enormous power of a massive black hole such as this is the energy source for quasars. Illustration above: On a smaller scale, black holes with thousands or millions of solar masses can tear apart a passing star. Most of the star material, however, would be flung into space rather than consumed by the hole.

Nearby galaxies are trapped, looping in vast orbits around the dominant ellipticals.

These emperor galaxies are cannibals, growing fat at the expense of other galaxies that are partially denuded or completely torn apart and swallowed up as they pass the giant ellipticals. Sometimes, more than 100 smaller galaxies swarm around a dominant giant elliptical, many of them barely more than the nuclei of once elegant spirals. In photographs, cannibal elliptical galaxies look like vastly overgrown globular clusters, but other instrumentation reveals that they are surrounded by enormous halos of stars and gas evidently ripped from passing galaxies. The halo acts as an effective net for ensnaring gas from more distant galaxies that pass by. The cannibal galaxy thus extends its influence and leaves its neighbor galaxies as star skeletons without nebulas to form new generations of stars. Meanwhile, the trapped gas slowly falls toward the cannibal, adding to its mass.

Some giant ellipticals have quasarlike jets reaching out from their nuclei to a distance equal to several times the galaxy's diameter. There may be a direct connection between that activity and quasars. During a violent encounter, when huge quantities of gas fall into the dominant elliptical from another galaxy, the big galaxy's central black hole might be fueled up to quasar levels. The jets could be evidence of this activity. The nearest dominant cannibal, known as M87, has just such a jet and appears to contain a central black hole of five billion solar masses, easily big enough for quasar-level bursts.

All this violence among and within galaxies is alien to our star city, the Milky Way. Our nucleus black hole is puny compared with the one in M87. And the closest we have come to a galaxy collision is the event that caused the disruption and recent burst of star formation in the Large and Small Magellanic Clouds, our satellite galaxies. Astronomers are uncertain about what happened, but a sideswipe involving these two small galaxies and the outer arms of the Milky Way probably triggered the activity. Sooner or later, all satellite galaxies are destined to be ripped apart or

UGC10214, one of the most spectacular examples of a gravitationally distorted galaxy, is about 90,000 light-years in diameter and 420 million light-years away, in the direction of the constellation Draco. This image was taken by the Hubble Space Telescope's Advanced Camera for Surveys, installed on the orbiting telescope in 2002. The power of this camera enables astronomers at the Hubble controls to peer billions of light-years past the main galaxy and see details in hundreds of background galaxies.

swallowed up or left totally devoid of gas and outrider stars. The Andromeda Galaxy's two small elliptical satellite galaxies may have suffered the latter fate. However, both Andromeda's companions and our satellite galaxies are too small to do major structural damage to the main galaxies.

The placid history of the Milky Way Galaxy seems largely the result of its location in a small, dispersed galaxy cluster called the Local Group, which lies on the outer fringes of the Virgo galaxy supercluster. (A galaxy supercluster is a collection of galaxy clusters that seem to be gravitationally bound to each other but do not merge.) The Local Group galaxies have been producing generation after generation of stars for billions of years without bothering their neighbors. But a local display of fireworks is on the distant horizon.

Sometime between three and five billion years from now, the Milky Way and Andromeda galaxies will either collide or have a disruptive close encounter. Astronomers can't as yet determine how close the two will come to each other or exactly what the fate of either galaxy will be. But from observations of other galaxies, once a close encounter occurs, a death spiral begins that ultimately results in a collision and merger, even though it may take billions of years.

Although planetary families like the solar system would probably survive a galactic merger unscathed, there are fundamental differences between spiral and elliptical galaxies that may have some bearing on the life question. Spiral-arm stars never venture close to the core. Our Sun has a near-circular orbit around the nucleus. An elliptical galaxy, however, is a tangle of stellar orbits. Most of its stars swing in looping paths that carry them within a few thousand light-years of the core.

Just how this would affect a planet of such a star is uncertain. Even though stars near the core of an elliptical galaxy are hundreds of times closer together than in our Sun's vicinity, a planet like Earth would not be disturbed from its solar orbit, but the occasional radiation bath near the elliptical's core might be a problem. The most we can say at this time is that ellipticals cannot be ruled out of the life game. But even if they were, there are billions of spiral galaxies in the universe.

The Universe Train

Much more of our journey has yet to unfold, but this is a good point to pause and regroup. Our solar system has faded behind us, and the universe of galaxies now stretches out in all directions. But at the same time, the true immensity of the cosmos strains comprehension. For example, the galaxy on the facing page is 420 million light-years away. A beam of light would require 90,000 years just to travel from one side of the galaxy to the other.

To keep things a bit more manageable, let's stay with one immensity at a time. For now—a demonstration of just the *number* of stars in the known universe.

Imagine a dump truck filled with construction sand. There are about 100 to 300 billion grains of sand in the truck, roughly the same as the number of stars in the Milky Way Galaxy. A thimbleful of sand scooped from the load—several thousand grains—represents all the stars visible to the unaided eye on a dark, moonless night.

But to imagine the universe of galaxies, think of a train carrying enough sand in hopper cars to represent all the stars in all the galaxies in the universe. For the cargo to correspond to the universe's estimated 10 billion trillion stars, the train would have to use all the sand on all the beaches on Earth and would stretch around the planet 25 times.

For the most vivid perspective, picture the Universe Train approaching a level railroad crossing. It rumbles by at a steady pace—freight-train speed—one hopper car per second. At that rate, it would take *three years* for the entire universe of sand-grain suns to pass.

At 60 million light-years, M87 is the nearest giant elliptical galaxy to Earth. This Hubble Space Telescope shot of the nucleus of M87 reveals a quasarlike jet emerging from a central black hole with an estimated mass of five billion Suns. Compare this image with the illustration on page 111.

INTO THE ABYSS

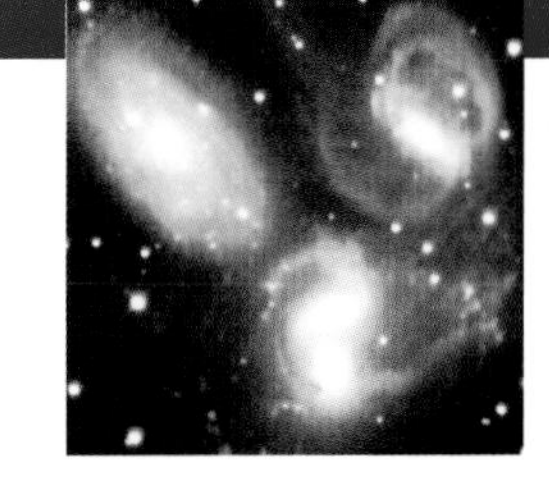

CHAPTER SEVEN

By human standards, one cubic light-year is an enormous chunk of space, enough to hold the Sun, its orbiting family of nine planets, a million roving asteroids and a trillion comets on giant looping paths which carry them so far out that Mother Sun is reduced to a bright star in the firmament. But on the scale of the known cosmos, this is merely our backyard—just one infinitesimally tiny pocket in a universe of a million trillion trillion cubic light-years.

Suppose we could select any one of those million trillion trillion cubic light-years as a site for exploration and a survey of surrounding space. How would the cosmic landscape vary from one random site to another? Ninety-nine times out of a hundred, an intrepid explorer willing to embark on such a blind voyage would emerge in a pristine vacuum embedded in total blackness. Absolutely nothing would be visible to the unaided eye. Resorting to binoculars, the traveler might spy a few smudges of light—some remote galaxies—but statistically, that would be unlikely too.

The universe is almost entirely empty space. Planets, stars and galaxies are scattered here and there in an abyss of nothingness (at least as far as human senses are concerned). There is one galaxy for every million trillion cubic light-years of space and one star for every billion cubic light-years. But they are not evenly distributed. Stars are swarmed in galaxies, and galaxies congregate in clusters. On the largest scale, the clusters of galaxies are themselves arrayed in larger groups known as superclusters, each separated by enormous voids. An observer plunked randomly in the universe would probably land somewhere in one of the voids that occupy most of the cosmos. In regions where galaxies exist, the superclusters are like archipelagos scattered across the cosmic ocean.

Surveys of galaxy distribution have uncovered a structure, perhaps a grand design, to the universe. The voids are vast, roughly spherical zones that can be several hundred million light-years across. The only substantial objects within these voids would be the occasional errant galaxy or globular cluster that long ago escaped the family gravitational attraction of the galaxy superclusters.

Around the "surfaces" of the voids lie the galaxy superclusters. The entire affair resembles the cellular structure of soap bubbles: Where two voids meet, a sheet of galaxies is likely; zones of multiple intersections seem to produce denser ribbons and tendrils of galaxies. The junctions of several void surfaces at one site can generate the most populous knots, marking the cores of galaxy superclusters.

Just as soap bubbles in a sink are mostly air with a little soap, so the universe on the whole is primarily vacant. However, layer upon layer of bubbles give the impression of substance, as do photographs crowded with galaxies.

Some astronomers suspect that the voids are the realm of galaxies that never underwent bursts of star formation. These ghost galaxies would have the mass of normal galaxies, with most of it in the form of gas. Evidence for this theory, however, remains weak. More likely, the voids are just what they appear to be—largely empty. They trace their origin to conditions in the early universe when slight differences in density seem to have marked out the regions of space where superclusters and voids would eventually form.

The universe's cellular anatomy is the most recent

Let your soul stand cool and composed before a million universes.

Walt Whitman

Facing page: This is what the universe really looks like—an ocean of black space speckled with galaxies of all shapes and sizes. The Hubble Space Telescope was aimed at an inconspicuous region in the constellation Fornax for this 275-hour exposure, known as the Hubble Ultra Deep Field, which shows galaxies up to 12 billion light-years away. The image covers an area of the sky no larger than a grain of sand held at arm's length. Above: Stephan's Quintet, a famous group of galaxies.

Woven across the cosmic landscape like the threads in some vast tapestry, galaxy superclusters define the ultimate structure of the universe. This image is a plot of nearly a million individual galaxies (as seen from Earth) and represents about half of the entire sky. Far from being scattered at random, galaxies are clustered, and the clusters congregate in superclusters that form the filaments seen here. The dense cluster at center is the Virgo galaxy supercluster. It stands out because our Local Group of galaxies—the Milky Way, Andromeda, Triangulum and others within a radius of a few million light-years—is on the edge of the Virgo group, making it the nearest cluster among the throngs shown here.

link in a chain of cosmological revelations that dates back to 1924, when Edwin Hubble used the Mount Wilson 100-inch telescope (then the world's largest) to examine individual stars in the Andromeda Galaxy. This was the first conclusive proof of the existence of galaxies other than the one we inhabit. For decades prior to Hubble's triumph, astronomers were uncertain about the nature of the spiral nebulas, as galaxies were then called. Before that, philosophers swung just as much weight as astronomers in discussions of the structure and extent of the cosmos. The discovery of the universe is almost entirely a 20th-century enterprise. Today, the uncharted territory is on more distant shores, where the fireball that gave birth to our universe may have created a multitude of other universes.

The Expanding Universe

In less than a minute, the universe will increase its volume by a trillion cubic light-years. Propelled by the force of its explosive birth 13.7 billion years ago (the best estimate of the universe's age as of 2004), the universe is expanding like an inflating balloon.

Albert Einstein mathematically predicted the expansion as a side effect of his theory of general relativity in 1916, but the idea was so revolutionary at the time that even Einstein balked at the implication of his own equations. In a move that he later called "the biggest mistake of my life," he inserted a term he called the cosmological constant in the key equation to keep the universe static. Meanwhile, the first evidence that the galaxies are receding from one another was quietly being gathered by astronomer Vesto Slipher at the Lowell Observatory in Flagstaff, Arizona, using the same telescope which had been built to study the "canals" of Mars. The canals proved to be optical illusions, but the receding galaxies are fact.

Slipher announced his findings before the exact nature of galaxies was understood, so it was not clear at the time exactly what these objects speeding away from Earth were. But there was no doubt about the fact itself. It hinged on analyses of the spectral "fingerprints" contained in the galaxy light. Those fingerprints are the positions of lines in the spectrum of starlight, caused by glowing gases such as hydrogen or by the absorption of starlight by gases. The lines are always in the same locations relative to one another, like the keys on a piano, but shifts in the entire set of lines reveal whether the galaxy is approaching or receding. The amount these lines are shifted away from the positions they occupy in the spectrum of a stationary laboratory source indicates a velocity in the observed object. It is like having two identical-looking piano keyboards, one in concert pitch but the other pitched a tone and a half lower, so the pianist has to play two keys to the right.

If the lines are shifted to the red, or longer, wavelength end of the spectrum, the source object is moving away from us. A blueshift means it is moving toward us. The effect is produced by the same principle that causes the pitch of a train horn to change as the train approaches the listener and then recedes: The wavelengths of sound are compressed to a higher pitch as the train draws nearer and are stretched to a lower pitch as it speeds away. Light displays similar wavelike characteristics to an observer studying approaching and receding celestial objects.

Slipher found that the vast majority of the galaxies he studied had a redshift, indicating the wavelengths of light were being stretched to lower frequencies as the galaxies receded. But he was pushing his 24-inch telescope to the limit. It was Edwin Hubble, using the much larger 100-inch telescope, who greatly broadened the galaxy redshift census. He made the crucial link, showing that not only are the galaxies moving away from us, but the farther away a galaxy is, the faster it is receding. He had discovered the expanding universe. It was a paradigm shift of monumental proportions.

On the face of it, the idea of an expanding universe suggests that everything is moving away from us, placing our galaxy at the center of the universe. Could this, in fact, be the case? And why do more distant galaxies move away faster?

A loaf of raisin bread provides a reasonably accurate demonstration of the phenomenon. When the loaf is being prepared for baking, the raisins are randomly distributed throughout the dough. As the dough rises, it expands fairly evenly to several times its original size. During the expansion, the raisins are carried along with the dough, each moving away from the others. Any one raisin will "see" all the remaining raisins receding from it, which creates the illusion that each raisin is at the center of the expansion. As the loaf doubles in size, a raisin one centimeter away will recede to two centimeters, while one that is two centimeters away will have moved to four centimeters. The more distant raisin therefore moves twice as far in the same time interval. Thus during the expansion, a raisin twice as far away recedes twice as fast.

This analogy, in which raisins are galaxies, deftly demonstrates both why we seem to be at the center of the expansion and why the farther away a galaxy is, the faster it seems to be receding.

The raisin-bread model portrays another aspect of the universe, in that it shows the raisins being carried by the dough rather than moving through it. The dough represents space in the real universe. Although space is, for practical purposes, a vacuum, it is part of the universe. A point in the universe exists in both space and time whether there is any matter there or not. Light and other forms of radiation propagate through space, and gravity holds sway throughout the universe (although the mechanism by which its force issues through space is still poorly understood). It is space that is expanding, carrying the galaxies along with it.

Galaxy clusters resist this trend, though. The gravitational grip of the cluster members on each other is usually strong enough to hold the group together. The superclusters, more loosely bound than the clusters within them, are also part of the expansion, but due to the mutual gravitational pull of the group, their constituent galaxy clusters drift apart at a slower rate than the overall expansion. The full expansion rate is observed only in the redshift of galaxies in remote superclusters receding from our supercluster.

But a problem developed in this scenario. In the 1970s, it became clear that galaxy clusters are not even close to being massive enough to produce the gravitational pull needed to hold themselves together. To keep from drifting apart, the clusters should contain more or larger galaxies. Yet the clusters are clearly gravitationally bound.

The mystery of the "missing mass," as it was called at the time, fully emerged when astronomers analyzed the motions of galaxies in clusters by measuring the differences in the redshifts of the cluster members, which allowed them to compare the velocities of individual members with the group average. The investigations revealed that member galaxies in the clusters were not flying away from each other as they should be if they were flowing with the overall expansion of the universe. It didn't add up.

The Dark Matter

The confirmation that there is more to the universe than meets the eye came when researchers took the first rigorous estimates of the total number of stars in each galaxy of a galaxy cluster. The discrepancy turned out to be huge. On average, galaxy clusters contain less than one-thirtieth the amount of material

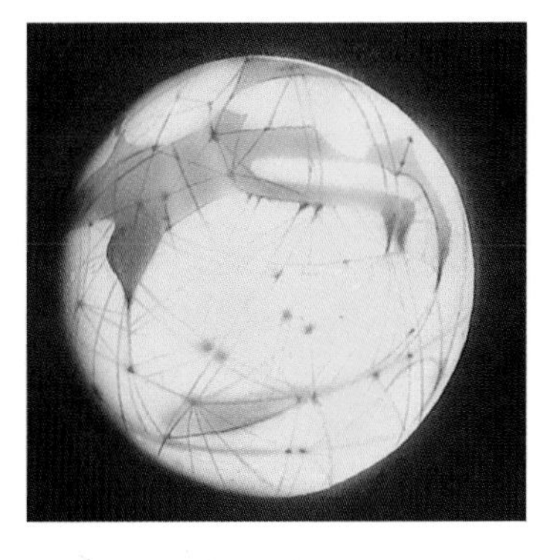

In one of the great ironies of astronomy, the telescope designed and built expressly to observe the "canals" of Mars, which proved to be nonexistent, ultimately revealed the first evidence that the universe is expanding.

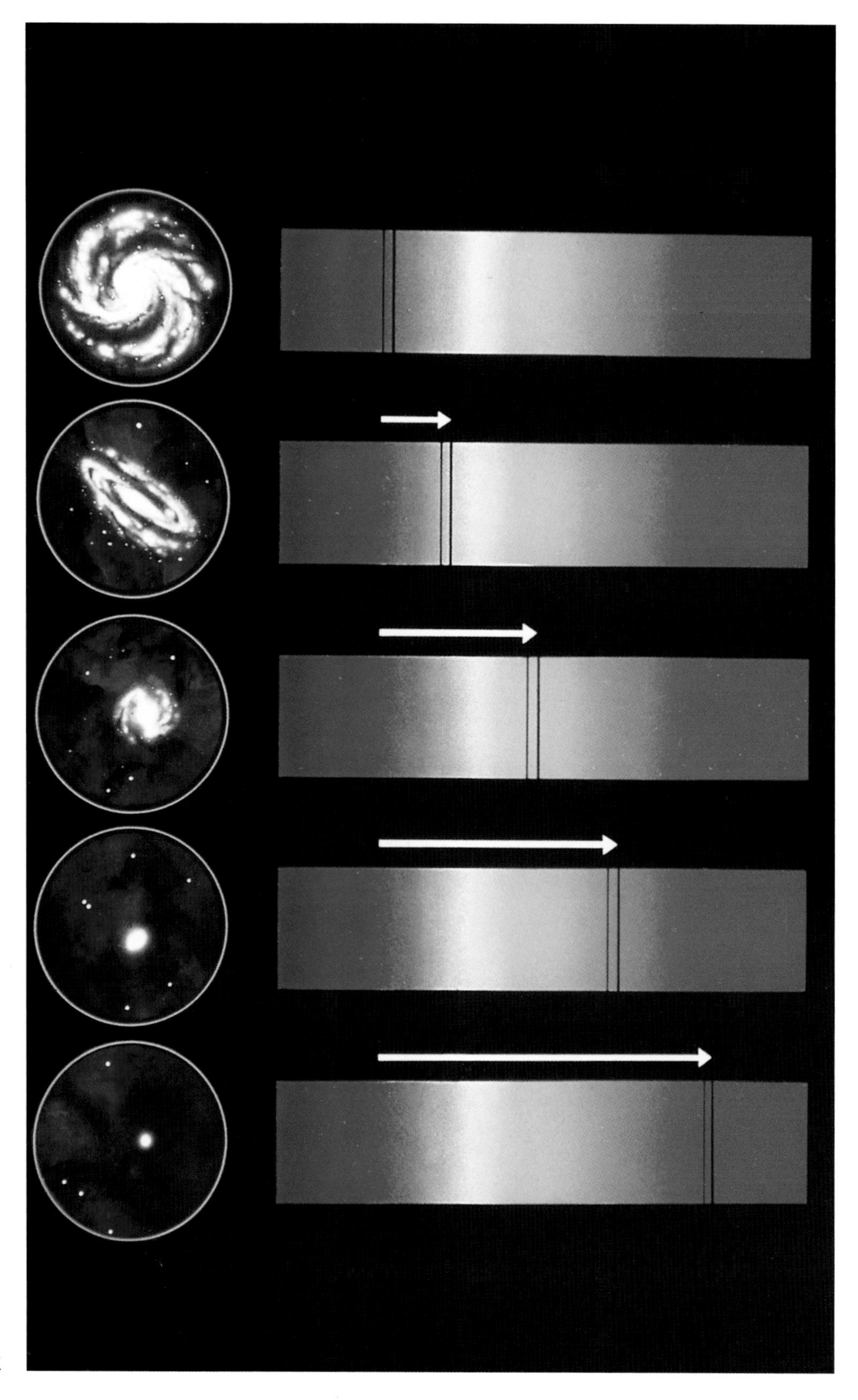

needed to keep the clusters together by gravity. Judging from the visible mass of their galaxies, the clusters should have dispersed long ago, leaving galaxies evenly spread across today's sky. But since galaxies are anything but evenly spread, some invisible "stuff" must be providing the gravity to hold the galaxy clusters and superclusters together against the force of the universe's expansion.

By the late 1970s, astronomers were convinced that the invisible "dark matter," as it was beginning to be called, must exist. The task then became identifying what it is or, failing that, *where* it is. Some of the matter clearly had to be inside galaxies, somewhere between the stars. Studies of motions of stars within our own galaxy suggested that a likely reservoir would be found beyond the Sun's distance from the hub of the Milky Way Galaxy.

The Sun takes 200 million years to complete one orbit around the galaxy's nucleus. Its orbital speed is determined by the amount of material between us and the center of the galaxy, which is about 100 billion times the mass of the Sun. So a star at the Sun's distance from the center of the galaxy (27,000 light-years) "feels" the gravitational influence of all the matter—stars, planets, dust and gas—between it and the galaxy's nucleus, and its orbital velocity is determined by that total mass.

Although the density of stars seems to drop off rapidly at the outer edges of the spiral arms—18,000 light-years beyond the Sun—the amount of mass out there does not. Rotation measures of other galaxies revealed the same thing, with most of the invisible mass in the outer galactic fringes.

The Andromeda Galaxy is a typical example of the enigma this presents. The combined light from all its stars is 25 billion times brighter than the Sun. Since the average star is substantially dimmer than the Sun, this figure translates into a total stellar mass of about 300 billion times the Sun's mass. And yet Andromeda's outlying objects are moving at velocities that indicate the galaxy is closer to several trillion solar masses. Almost every galaxy is similarly overweight, including our own. But the extra mass remains largely unseen.

There are only a limited number of conventional explanations. Stars produce light, and the light can be measured, so there cannot be more normal stars than we see. Then there are the unusual stars—stellar cadavers such as black holes, neutron stars, brown dwarfs and white dwarfs—none of which give off much light. However, a census of the region within a few dozen light-years of the Sun has not turned up even a small fraction of the number of white dwarfs that would be required (they would have to be far more numerous than normal stars to account for the invisible mass). Black holes, neutron stars and brown dwarfs could more easily remain undetected and should not be ruled out, but all investigations so far suggest that these objects account for, at most, a total equal to half the existing ordinary stars.

Could the invisible material be compressed into a supermassive black hole at the galactic core? Such objects do exist, with masses of several million suns for an average galaxy, but hundreds of billions of solar masses would be needed right in our own galaxy's core. And, in any case, most of the galactic mass cannot be a point source, because the orbital velocities of the Milky Way Galaxy's stars indicate that the dark matter is in the halo, not the core.

Huge numbers of pint-sized bodies—planets, asteroids, comets—scattered like trillions of motes of dust throughout our galaxy and others have been proposed to explain the dark matter. But modern theories of star formation simply cannot account for the vast number of these objects necessary to make up our galaxy's invisible deficit (for example, a million bodies the mass of Earth for *each* star in the galaxy would barely be enough). Theoretically, the creation of so much debris is unlikely without stars emerging with it.

The only definite dent in the invisible-mass enigma was the x-ray-satellite discovery of a thin, hot gas between galaxies in large, dense clusters. While the gas may be thin—a near vacuum by human standards—there is so much of it that it doubles the cluster's mass. Still, that's not nearly enough.

At the close of the 20th century, astronomers had been wrestling with the puzzle of the dark matter for decades. As we shall see, the story has some twists and turns yet to play out.

The Big Bang

Stepping back to the 1930s for a moment, imagine astronomers first digesting the stunning fact that the galaxies are speeding away from each other, then facing the startling realization that the universe must have had a beginning. The expansion, it was reasoned, was generated by a creation event—prosaically dubbed the Big Bang—whose multibillion-degree fury created the fabric of expanding space that is hurling matter in all directions. The effects of the genesis explosion can be seen today as the galaxy superclusters dash away from one another.

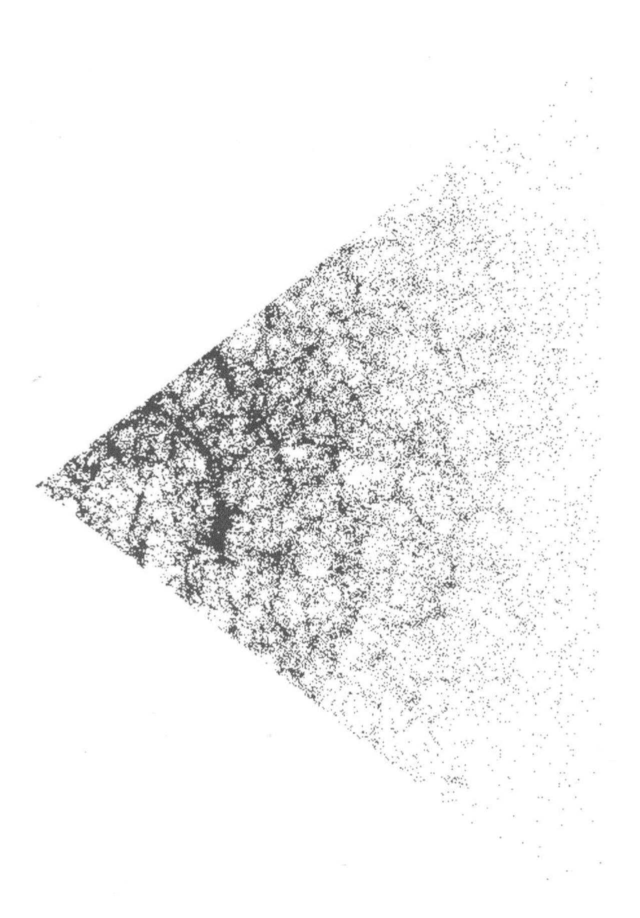

Left: A massive survey to determine the positions and distances of 220,000 galaxies in a wedge of the cosmos reaching out two million light-years was completed in 2002. Earth is at the point of the wedge; the most distant galaxies are to the right. Each dot represents at least several galaxies. Giant clusters of galaxies as well as voids containing almost no galaxies are prominently visible.
Facing page: The distance to a galaxy can be gauged by its redshift. A spectrograph captures telltale absorption lines (dark) or emission lines (light; not shown) in a spectrum of a galaxy. The farther the lines are displaced to the red, the faster the galaxy is receding and the more remote it must be—a relationship caused by the expanding universe.

But, some astronomers asked, could it be an illusion? Could the redshifts be caused by some phenomenon other than fleeing galaxies? They wanted proof that the Big Bang happened. The debate peaked in the 1950s with the introduction of the steady state theory, which suggested that the expansion is caused by the *continuous creation* of matter from the vacuum of space. The steady state theory's fundamental assumption is that there was no beginning to the universe, nor will there be an end.

Questions of origins can be probed because, on the scale of the universe, distance is a time machine. The more distant something is, the longer its radiation takes to get here and the younger it appears. Therefore, if astronomers can look deep enough, the heat of the genesis fireball that preceded the origin of the galaxies should be detectable. It would appear in all directions, because everything emerged from it, so any sight line would lead to the creation event's signature.

Princeton physicist Robert Dicke labored for years to determine exactly what the Big Bang would look

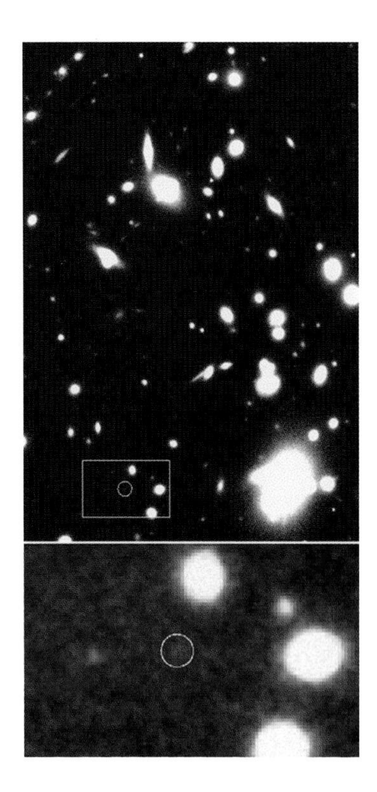

Above: The most distant galaxy known is 13.2 billion light-years away, according to redshift measurements. The record-holding galaxy is silhouetted against the glow of the cosmic background radiation in this false-color infrared image, bottom. Right: Three pairs of Hubble Space Telescope images show three distant galaxies before and after a supernova explosion. These are type Ia supernovas, studies of which led to the discovery of the dark energy in 1998. In some cases, a supernova can exceed the entire light output of all the stars in its host galaxy.

like. He calculated that we might still detect the embers of the creation blast vastly redshifted by the expansion of the universe over the intervening aeons. Since temperature is related to wavelength, Dicke said, the redshift would be so great, the radiation would not be visible light but, rather, very cool and invisible microwave radiation. This is comparable with the train in our redshift analogy receding at such speed that the sound of its horn is below audible levels.

Unaware of Dicke's work, Arno Penzias and Robert Wilson of Bell Laboratories in Holmdel, New Jersey, were testing a new microwave antenna in the spring of 1965 when they noticed a faint but persistent signal coming from all directions. It was exactly what Dicke had predicted. It's now called the three-degree cosmic background radiation, because that is how cool the redshifted creation fires appear (three Celsius degrees above absolute zero). This is the most powerful evidence that the Big Bang happened. In effect, the echo of the creation event can still be heard some 13.7 billion years later.

Origins

At the moment of creation, according to current versions of the Big Bang explanation for the existence of our universe, everything we know and see today started from a point no bigger than a proton. Space and time were born at that instant. As time grew from instant zero, space grew from size zero. As space expanded, there was more room in the dense, intensely hot sea of energy, and a steady cooling ensued. Eventually, space expanded and cooled enough for atoms to form without immediately breaking apart. About 380,000 years after the Big Bang, matter in the universe was spread out enough for light and other forms of radiation to flow through space without being quickly absorbed by jostling subatomic particles. That radiation is still flowing through the universe today, detectable as the three-degree cosmic background radiation.

The background radiation is the oldest thing we can observe. In 1992, astronomers announced that NASA's Cosmic Background Explorer satellite had detected slight variations in the temperature of the background radiation as it reaches us from different directions. The variations are tiny, just 3/100,000 Celsius degree from place to place, but they indicate that way back then, matter was not spread evenly over the universe. The slightly denser regions appear to be a bit cooler than the less dense regions.

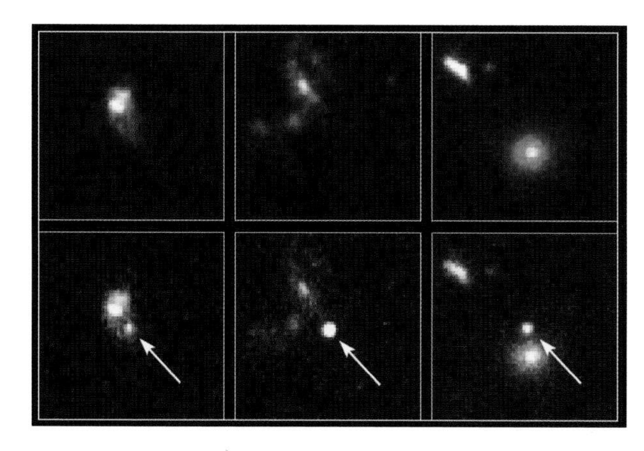

This discovery showed, for the first time, that matter was starting to collect in patches even back near the beginning of time. Astronomers think that those tiny temperature ripples were the seeds for the vast clusters of galaxies which would arise 500 million to one billion years later. If the temperature ripples had not been found, the emergence of galaxies from the Big Bang would be tough to explain and the Big Bang theory might have been seriously undermined.

Although not entirely without its problems, the Big Bang theory is, at the moment, by far the best explanation we have for the universe's existence. Competing theories, such as steady state, the plasma-energy universe and other lesser-known hypotheses, are championed by a tiny minority of cosmologists—fewer every year, it seems. The Big Bang is mainstream not because it appeals to some Judeo-Christian subtext of creation and a creator; scientists accept it as conventional wisdom because observations show that it best explains the universe we see around us.

Cosmic Inflation

More and more in cosmology, an understanding of the very large—the workings of the entire universe—has also meant a comprehension of the very small. Beginning around 1980, investigations of the interaction of the smallest subdivisions of matter, a field known as quantum physics, began to have explicit applications to astronomy, especially in connection with what caused the Big Bang.

Quantum physics had previously been used to predict that a fraction of a second after the genesis instant, the universe was in an unimaginably compressed state, confined to a volume the size of a baseball. In this state of compression, an object the mass of our Sun would have been the size of an

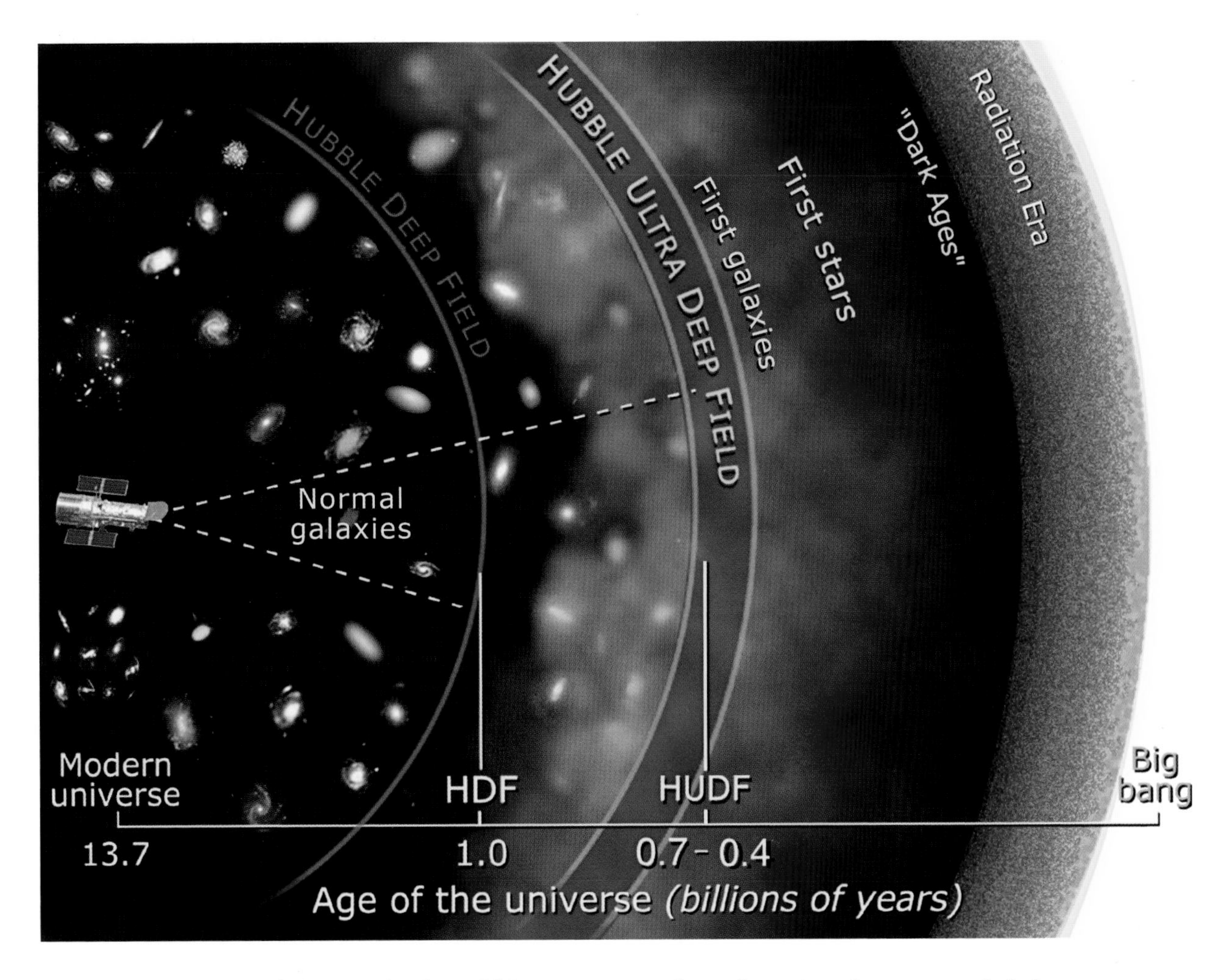

This simplified illustration shows the power of the Hubble Space Telescope when it was equipped with its first camera used for the Hubble Deep Field (HDF) and, after 2002, when it received its upgraded camera used for the Hubble Ultra Deep Field (HUDF), which is reproduced in four sections in this book (pages 19, 118 and 132-33). The farther a telescope can peer into the depths of space, the farther it sees into the past. Of course, the Hubble and other telescopes record galaxies at all distances, from near to far, in the same image. Only after a galaxy's redshift is measured can astronomers estimate its distance.

atom and something as dense as lead would be, comparatively, the thinnest of gases. It was a seething cauldron of quarks, leptons and gluons, at a temperature 100 billion trillion times higher than that inside an exploding thermonuclear bomb. From this primitive state, all matter and energy we perceive today emerged. But that is where the conjecture stopped. Answers to questions addressing just where the ingredients for the primordial energy ball initially came from seemed totally inaccessible.

However, new developments in quantum-physics theory have made it possible to trace events back to a point only one trillion trillion trillionth of a second after the creation, when the universe was smaller than a proton. Temperatures and pressures were so prodigious that the four fundamental forces of nature—the electromagnetic force, the weak and the strong nuclear forces and gravity—were unified with equal and indistinguishable strengths. It was a time of supreme elegance and ultimate simplicity.

As the embryonic universe expanded, the temperature dropped, permitting gravity to release itself from the other three forces. The expansion continued for another billionth of a trillionth of a trillionth of a second until the next transition occurred: the separation of the strong nuclear force.

That event, known to physicists as symmetry breaking, can be thought of as a change of state, something like the boiling-point transition from liquid water to gaseous water vapor. Although the temperature change is slight, there is a huge expansion in volume when the water makes the transition from a liquid to a gas. Similarly, the energy of symmetry breaking in the primal universe resulted in an instantaneous injection of a vast amount of expansion energy that drastically enlarged the universe. The primordial cosmos "hyperinflated" to at least a billion trillion times its former size—a greater increase than a grain of sand expanding to the size of Earth.

Called the inflationary-universe scenario, the

EVOLUTION OF THE UNIVERSE

From Time Zero to the Present

Quantum space-time "foam" from which the universe emerged in the Big Bang. Extreme conditions beyond present theory, because all forces of nature indistinguishable from one another.
Time: zero.

Gravity separates from the unified forces.
Time: 10^{-43} to 10^{-35} seconds after the Big Bang.
Temperature: one million trillion trillion degrees C.

Cosmic Inflation
Time: 10^{-35} to 10^{-33} seconds.
"Observable" universe (a tiny part of total inflationary cosmos) expands from one-trillionth the size of a proton to the size of a baseball.

Radiation Era
Time: 10^{-33} seconds to 380,000 years.
Three other fundamental forces of nature—strong and weak nuclear forces and electromagnetism—separate from being unified (10^{-28} to 10^{-12} seconds). Universe continues to cool as it expands; quarks, then protons and neutrons form. At 380,000 years, the universe has cooled enough (60,000 degrees C) for atoms to remain stable.

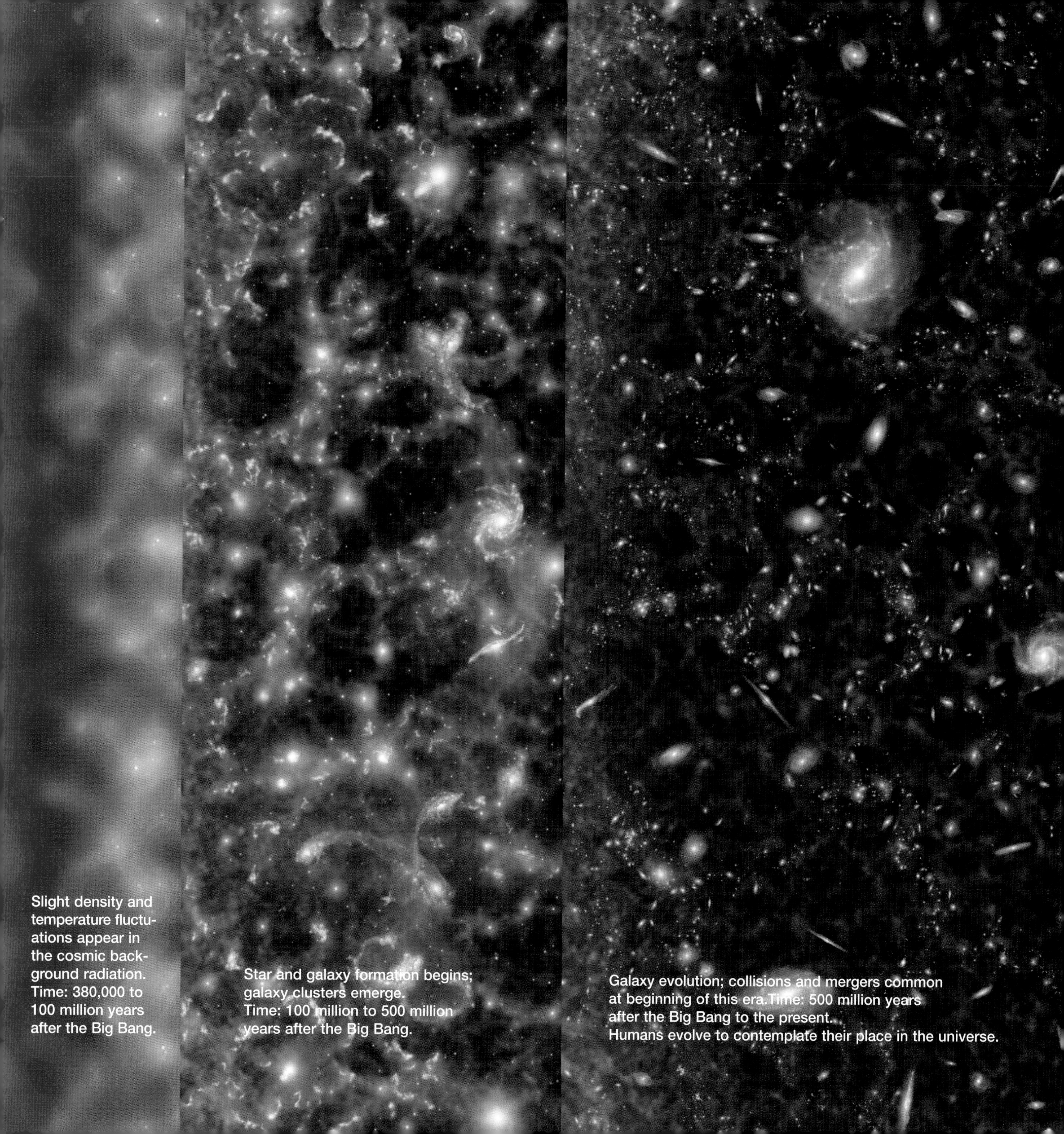

Slight density and temperature fluctuations appear in the cosmic background radiation. Time: 380,000 to 100 million years after the Big Bang.

Star and galaxy formation begins; galaxy clusters emerge. Time: 100 million to 500 million years after the Big Bang.

Galaxy evolution; collisions and mergers common at beginning of this era.Time: 500 million years after the Big Bang to the present. Humans evolve to contemplate their place in the universe.

These false-color images represent the entire sky spread out flat like a world map. The colors are slight temperature variations in the cosmic background radiation as measured by the Cosmic Background Explorer satellite in 1992, top, and by the much more detailed Wilkinson Microwave Anisotropy Probe in 2003, bottom. The temperature ripples (pink slightly warmer than blue) are thought to be the primordial seeds for galaxy superclusters that arose 500 million to one billion years later.

theory includes the astounding prediction that within the supercooled, hyperinflated cosmic crucible, a nearly infinite number of universes bubbled into existence like foam, stabilizing the structure. Our baseball-sized universe, whose origin seemed so mysterious a few years before the inflationary-universe theory, was one of them. From that point until today, about 13.7 billion years, each of those universes—ours among them—has been evolving. The expanding universe of stars and galaxies that we inhabit may be but a tiny segment of the whole creation, an atom-like blip in the body of a colossal cosmic realm.

Some of these universes could be similar to our own; others might be composed of antimatter (where positive and negative charges of subatomic particles are opposite to those in our universe). Still others may be ruled by laws with fundamental constants differing from those which govern the forces of nature in our universe. For example, gravity could be weaker or stronger relative to the other forces. A universe under such alien conditions might disperse its matter so rapidly that stars and planets could not form. Conversely, a slowly expanding universe would have collapsed back on itself long ago, producing a maelstrom of massive black holes. (Although inflation seems to defy the theory of general relativity by exceeding the velocity of light, it does not. The part of the expanding volume that became our universe never broke the speed limit. Because each universe is disconnected from all others, relativity is not violated.)

When we thought our universe was all there was, it was impossible to explain why it has the properties it does—properties that specifically allow stars and galaxies to form and matter to arrange itself into planets and life. Now, with practically an infinite number of universes emerging from the inflationary-universe theory, there can be a universe with every combination of properties imaginable. A universe like ours is but one of them. Just as Earth is teeming with life while the vast majority of worlds are thought to be unsuited for the emergence of biologies, our universe has stars, galaxies, planets and a unique balance of physical properties that is unlikely to exist in many other universes.

Scientists have known for some time that subatomic particles called virtual particles materialize from tiny fluctuations in a vacuum, then vanish. Although the process normally occurs in less than a trillionth of a second, it has been observed and can be accounted for in quantum-physics theory. Since the appearance of virtual particles is a random process, it is conceivable that given an infinite amount of time, one of these vacuum fluctuations with exactly zero energy would occur. Such a fluctuation would not vanish after the normal ephemeral lifetime of a virtual particle. Instead, it would absorb almost unlimited energy from the vacuum, as if a hole had been punched in a dam, and would emerge as a colossal fireball that would evolve just as the inflationary-universe theory predicts. Astonishingly, the entire cosmos is, in essence, a reexpression of the primal vacuum that predates our own universe.

The new concepts have a further appeal. They are an extension of a trend which began in the 16th century, when Copernicus suggested that the Sun, not Earth, is the center of the cosmos. Since then, our Sun has proved to be just one among hundreds of billions of stars in the Milky Way. In turn, our galaxy is but one among billions in the observable universe. Now, our universe is seen as merely one among a multitude of cosmic bubbles drifting in a far greater ocean, the universe of universes.

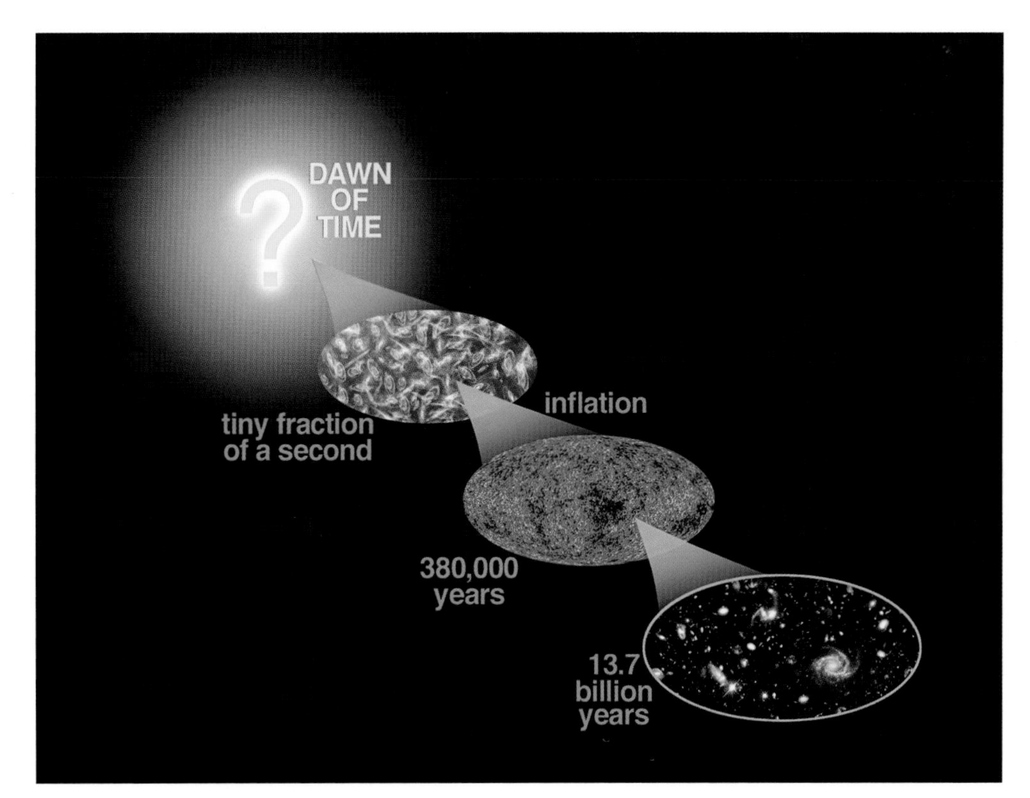

This NASA illustration depicts the first light to break free in the infant universe, 380,000 years after the Big Bang. Patterns imprinted on this light (shown in more detail on opposite page) reflect the conditions set in motion a tiny fraction of a second after the Big Bang. In turn, the patterns are the seeds for the development of the clusters of galaxies we now see billions of years later.

Vast Invisible Universe

British geneticist J.B.S. Haldane once said, "The universe is not only queerer than we imagine, it is queerer than we *can* imagine." Although Haldane uttered his quip half a century ago, it has been dramatically reinforced by discoveries in cosmology. Foremost among recent findings is persuasive evidence that more than 99 percent of the universe is not stars or galaxies but is, instead, invisible. And not only is most of it invisible, its nature is unknown.

The invisible stuff comes in two flavors, each completely different from the other. One we have already discussed—the dark matter in and around galaxies that makes them act more massive than they appear. The dark matter is probably exotic particles not yet discovered in any physics lab. The current favorite particle is the neutralino, predicted to exist in abundance since immediately after the Big Bang but as yet unseen. The universe's cargo of neutralinos could outweigh the stars in the galaxies by a factor of 30, enough to account for the bulk of the dark matter. But all remains theory until particle detectors provide proof that they are the culprits.

But it was the discovery of the dark energy, something altogether different from dark matter, that puts

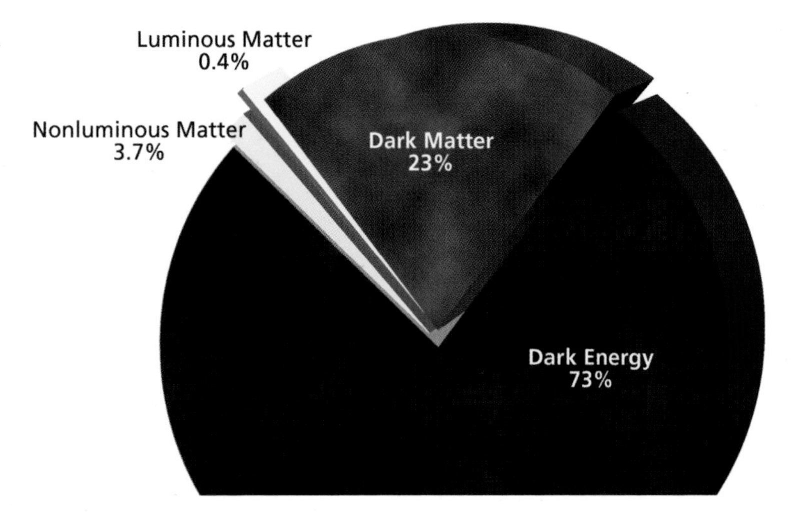

Discoveries at the dawn of the 21st century have provided convincing evidence that the part of the universe with which we are most familiar—asteroids, moons, planets, stars, nebulas and galaxies—accounts for less than 1 percent of the universe's total contents. Nonluminous matter (mostly hydrogen dispersed throughout the space between the galaxies) is about 3 percent of the contents. Familiar matter is thus a minor component of the universe. Though completely invisible, dark matter (formerly called the "missing mass") keeps galaxies bound into clusters. Dark energy, the largest and most mysterious component, acts as a repulsive force over vast distances and seems to be accelerating the expansion of the universe. (Total indicated comes to more than 100 percent because of rounding.)

Haldane's prophetic statement up in lights. The reality is that the galaxies we have come to think of as "the universe" are, in effect, a minor component suspended in an ocean of the two kinds of invisible stuff—roughly one-quarter dark matter and three-quarters dark energy.

The existence of dark matter has been evident since the 1980s, so dark matter has a (sort of) groping familiarity by now. It acts like normal matter in that more of it collects in some places than in others. Indeed, it was probably concentrations of dark matter that caused galaxies and galaxy clusters to form in the first place.

From Edwin Hubble's time until 1998, everybody assumed that the gravity of the collective mass of all the stuff in the universe—galaxies, nebulas, gas, black holes, dark matter—was enough to slow down the expansion. There seemed to be lots of direct and indirect evidence to support this theory. It was especially plain that gravitational attraction from dark matter prevents galaxy clusters like the one pictured opposite from breaking apart. It was a simple extrapolation to assume that dark matter, with perhaps some help from something yet to be discovered, was slowly decelerating the general expansion of the universe which began with the Big Bang 13.7 billion years ago.

Wrong.

The Accelerating Universe

The breakthrough year was 1998. Astronomers making redshift distance measurements of type Ia supernovas in remote galaxies had noticed that the maximum brightnesses of these exploding stars appeared dimmer than they should be at the measured redshift distances. Since all type Ia supernovas reach the same intrinsic maximum brightness, the results were unexpected and surprising.

These measurements seemed to indicate that the universe's expansion is accelerating rather than slowing down. In an accelerating universe, space would have been expanding more slowly in the past, stretching light waves more slowly and thereby showing less redshift than a decelerating universe. That's exactly what the new observations suggested: We live in a universe containing energy in "empty" space that is accelerating not decelerating the expansion of the universe. Astonishingly, but unmistakably, something is pushing the galaxies apart. As *Scientific American* magazine put it, "For cosmologists, it was like stepping on the brakes and feeling the car speed up, an exhilarating but disconcerting sensation that something wasn't working quite as it should."

A repulsive force that works only at vast distances is apparently intertwined with the fabric of space. Astronomers now call it dark energy.

Even though no one has any idea what dark energy is, its effects are real enough. Additional research reported in 2003 appears to have convinced even reluctant astronomers that dark energy exists and is the dominant player in the universe—the stuff that will ultimately control the fate of all we know. Not only don't we know what it is, we didn't even know it existed until almost 70 years after the discovery of the expanding universe! Haldane strikes again.

How did we reach the doorstep of the 21st century without realizing this stunning reality? One reason is that astronomers have spent all but the past few decades trying to understand what they could see. Only recently have they had the tools to explore the realm of the invisible. But the main reason is that the biggest clue to the full table of contents for the universe was buried inside something totally unexpected.

Hubble Perspective

Edwin Hubble (1889-1953) is generally regarded as the 20th century's greatest astronomer for his discovery of the expanding universe. By all accounts, Hubble was an imposing man. Tall and athletic, with chiseled good looks (picture Clint Eastwood or Gary Cooper, and you get the idea), he was not exactly the stereotypical image of a 1920s academic. Considering the reverential stature Hubble now holds in the annals of science, he remains one of the most enigmatic personalities in the history of astronomy.

After attending Oxford University as a Rhodes scholar, Hubble spoke with an Oxford accent for the rest of his life, yet he grew up in Missouri and, other than the stint at Oxford, lived in America his whole life. Aloof and distant, Hubble rarely socialized with the astronomers with whom he worked, preferring, instead, the company of the intellectuals who gravitated to Hollywood in the 1930s and 1940s. A photo from that time shows Hubble in casual conversation with zoologist and science popularizer Julian Huxley and Walt Disney.

The Hollywood connection, although odd to be sure, is not as bizarre as it might seem. The world's largest telescope at the time, the 100-inch reflector on Mount Wilson, was within driving distance of Los Angeles. Using this telescope in the late 1920s, Hubble measured the redshift of dozens of galaxies and proved, beyond doubt, that galaxies are moving not only away from us but away from each other in an expanding universe.

From that time until the 1990s, the assumption was that the combined gravitational attraction of all the galaxies, along with whatever other stuff the cosmos contains, would decelerate the expansion. The only question was whether the deceleration would be sufficient to bring the expansion to an eventual halt.

Now nature has turned the conventional wisdom on its head. The dark energy is not exactly an antigravity force. Instead of weakening with distance, as gravity does, it gains strength. Only at distances of billions of light-years do the effects of the dark energy dominate, which, in fairness to struggling researchers, explains why it remained undetected for so long.

On the positive side, once cosmologists realized what they were dealing with, the dark energy explains several long-standing theoretical questions about the overall picture of the universe. Chief among these is that the dark energy—whatever it is—seems to be a vital stabilizing force in the universe. Dark energy is, in effect, a balancing act, with gravity and matter providing one side of the equation and dark energy the other.

The picture that emerges is one in which the universe is made up of 73 percent dark energy, 23 percent dark matter, 3.7 percent nonluminous normal matter (mostly in the form of thin hydrogen gas dispersed between galaxies and in galaxy clusters) and 0.4 percent visible matter—stars, galaxies and the nebulas in galaxies. So the universe we see is less than 1 percent of all that exists.

With the dark energy's existence now widely accepted, the focus of cosmological research is to try to explain what dark energy is. The leading theory is that the vacuum of space itself contains the mysterious antigravity force which becomes apparent only over vast distances of billions of light-years. This idea goes back to Albert Einstein, who inserted such a force in his equations to prevent the universe from expanding or contracting. He called it the cosmological constant. Like everyone else at that time, he thought the universe was static, neither expanding nor contracting. Once he learned that the universe is, in fact, expanding, Einstein abandoned the cosmological constant, thinking it was unnecessary.

Today, Einstein would be smiling at the irony that even when he thought he was wrong, he was actually explaining one of the great mysteries of cosmology, 80 years ahead of his time. The newest research in this area suggests that the dark energy must be a force built

Above: A dense cluster of giant elliptical galaxies known as Abell 2218 gravitationally warps the light from more distant unseen galaxies into huge arcs. Known as gravitational lensing, the effect was predicted by Einstein. The amount of lensing seen here indicates that hundreds of trillions of solar masses of dark matter infuse and surround the enormous cluster.

On March 9, 2004, astronomers at the Space Telescope Science Institute unveiled the deepest portrait of the visible universe ever achieved. Called the Hubble Ultra Deep Field (HUDF), the 275-hour exposure reveals the most remote galaxies yet seen, roughly 13 billion light-years away. Each of the specks in this image is a galaxy of approximately 100 billion stars. The nearest galaxy in the picture is about one billion light-years distant. All the other objects are galaxies at some intermediate distance. (Exception: The objects with four symmetrical spikes are stars in our own Milky Way Galaxy.) The HUDF image reveals galaxies that are too faint to be seen by ground-based telescopes or even in Hubble's previous faraway looks, called the Hubble Deep Fields (HDFs), taken in 1995 and 1998. The HUDF field contains an estimated 10,000 galaxies but covers a patch of sky in the constellation Fornax smaller than a pinhead held at arm's length. In addition to the image's rich harvest of classic spiral and elliptical galaxies, a zoo of odd-shaped galaxies litters the field. These galaxies chronicle a period when the universe was younger and more chaotic. Photons of light from the very faintest objects seen here arrived at a trickle of one photon per minute, compared with millions of photons per minute from nearer galaxies. To observe the entire sky at this resolution with the Hubble Space Telescope would require almost one million years of uninterrupted observing.

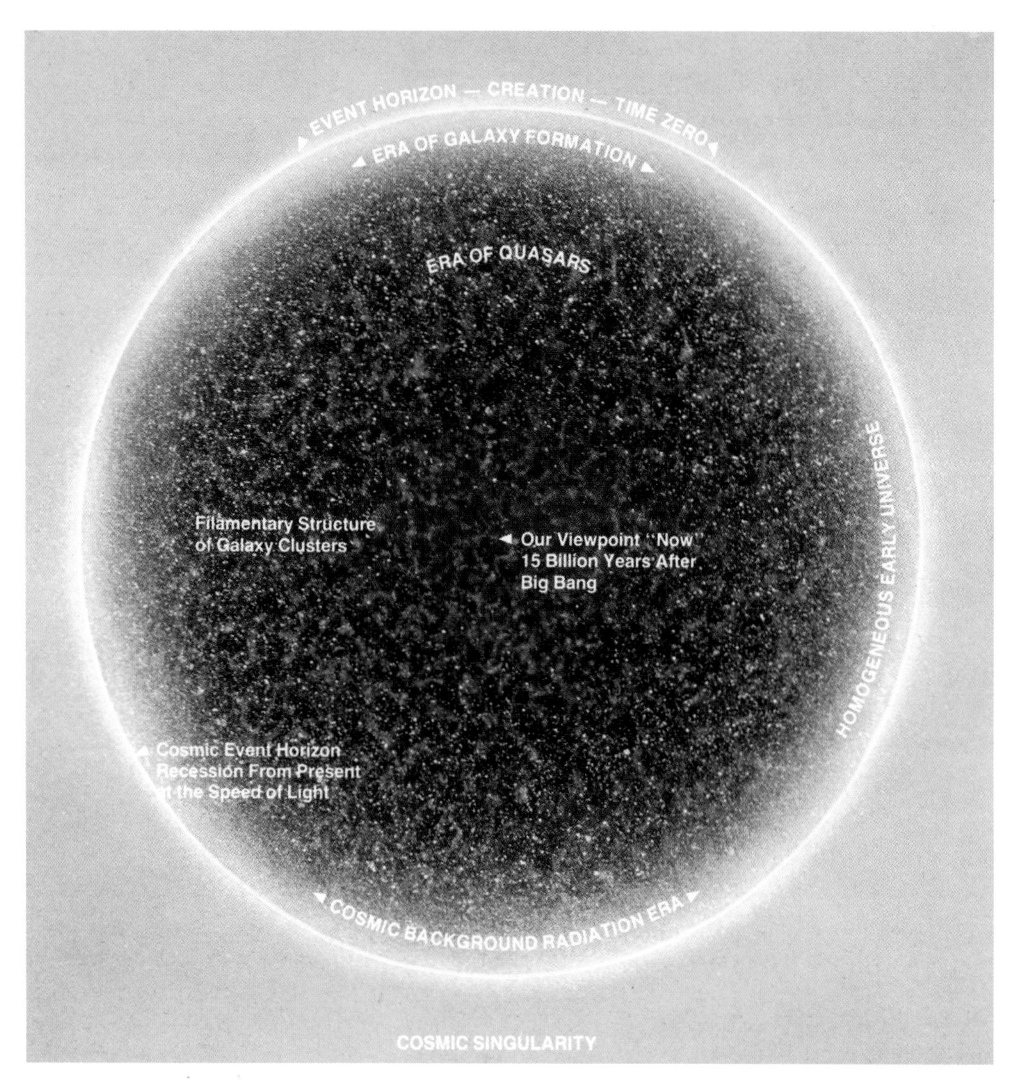

Although we seem to be at the center of the universe, it's an illusion. The universe is seen as it once was, not as it is today. Only events in our immediate vicinity are happening now; the rest are ghostly relics of the past. When we peer out to the greatest distances possible, we reach back in time to very near the beginning, just 300,000 years after the Big Bang, when the cosmic background radiation was released. To our perception, the genesis explosion is not a point in space but is everywhere. It defines the edge of space.

into space-time itself—essentially the same thing Einstein thought his cosmological constant must be before he abandoned the idea.

Tidying Up the Cosmos

The five years following the discovery of the dark energy proved to be a period of consolidation, a tying up of loose ends. Foremost among these loose ends was the estimated age of the universe at 9 billion years. Since stellar astronomers had measured the age of the oldest-known stars at about 13 billion years, a 9-billion-year-old universe was absurd. It was analogous to insisting a mother was younger than her children.

Since the 9-billion-year-age estimate was made before the discovery of the dark energy, cosmologists recalculated the universe's age, taking into account the weird properties of dark-energy acceleration. With acceleration counteracting the gravitational deceleration, the universe must have taken longer to reach its present state of expansion. When the adjustments were factored in, Bingo! The discrepancy disappeared. An expanding, accelerating, dark-matter, dark-energy universe with the measured properties our universe has should be 13.5 to 14 billion years old—exactly what the age of the stars would suggest.

In 2003, the Wilkinson Microwave Anisotropy Probe results (page 128) nailed down the age of the universe to 13.7 billion years, plus or minus 1 percent. As recently as the 1980s, estimates of the universe's age had ranged from 8 to 18 billion years.

Returning to the dark energy in a different context, however, we are not much closer to understanding what it is than was Einstein when he first suggested its existence as the cosmological constant —basically a stabilizing force in the cosmos. But current research gives cause for hope that the dark energy may reveal its first big secret.

Because the dark energy works in the opposite fashion to gravity (i.e., the larger the distance, the greater the dark energy's power), the more the universe expands, the more dominating the dark energy becomes. It already holds the premier place in the power hierarchy of the universe. But in the past, when the universe was smaller, there must have been a crossover point where dark energy lost its title to dark matter. Theorists think it was about five billion years ago. Although it will be a tall order for present telescopes to study objects five billion light-years away to examine this era, next-generation telescopes should be able to do it. Knowing how the dark energy has evolved over time will take us a long way toward understanding this most mysterious force in the universe.

Some researchers have suggested that the dark energy may be a manifestation of string theory, an idea which has been around in one form or another since the late 1960s, when it was first proposed by theoretical physicist Gabriele Veneziano. In its updated form, string theory at least gets a sympathetic hearing from many physicists. Although it is beyond the scope of this book to delve deeply into string theory, I can recommend two superbly written popular books on the subject by Columbia University physicist Brian Greene (see Further Reading). But there is one aspect of string theory I must mention. It deals with the most frequently asked question from audiences at the conclusion of a cosmology lecture: "Yes, Dr. Knowitall, but what came before the Big Bang?"

Many universes may have come into existence at the same time that ours emerged from the Big Bang. Since we cannot directly observe past the beginning, or "outside" our own universe, the existence of other universes remains theoretical. For the same reason, our universe could not be observed from the outside, as depicted here. Nevertheless, this illustration demonstrates some cosmological concepts. The sphere at lower right, almost out of the scene, is our present universe of galaxy clusters. Its predecessor is the universe at five billion years, when galactic collisions and quasars were more common. Other spheres show earlier epochs back to the Big Bang.

The moment of the Big Bang, illustrated at the extreme left on pages 126-27, is described as "beyond present theory," which is another way of saying, come back in 30 years, and we'll tell you. However, the basic idea of string theory is that elementary particles are not pointlike but, rather, infinitely thin, one-dimensional objects—strings. The advantage is that in a string-theory universe, the Big Bang still occurred, but because it did not involve a singularity—a moment of infinite density—the universe may have preexisted. String theory allows the universe to begin almost empty and build up to the Big Bang. Stay tuned.

The Cosmology Construction Site

Finally, a few words about cosmology as reported in newspaper science pages, news magazines and other mainstream news media. P.J.E. Peebles of Princeton University, a highly regarded elder statesman in cosmology, often sits on discussion panels at cosmology conferences as a cautionary "umpire." He made this comment in a public lecture:

"Research is a complex and messy business. I always feel uneasy when I see newspaper or magazine articles based on interviews with just one scientist. Even the most experienced researcher finds it hard to keep everything in perspective. How do I know how well this particular individual has managed it? I feel better when I see that the journalist has consulted a cross section of scientists."

Peebles says it is useful to think of our current ideas in cosmology as a construction site. Parts of the foundation are very solid, other parts less so. None of the workers have the final plans, though they feel a sense of special accomplishment when they contribute a room or hallway that stands the test of time. Others are kept busy reconstructing a tilting stairway or crumbling wall. Around the site, workers can be observed spending a lot of time thinking and dreaming about how the final edifice might look.

This chapter can be thought of as a State of the Universe Report—a summary of a work in progress. We live in the golden age of cosmology. Long may it last.

IN SEARCH OF EXTRATERRESTRIALS

CHAPTER EIGHT

Life may exist in yonder dark, but it will not wear the shape of man.

Loren Eiseley
1957

The newspaper cartoon strip *Frank & Ernest* sometimes features a pair of curious aliens scouting Earth in a flying saucer. One such excursion involved a decision about where to land. Finally, Alien A says to Alien B (they both look like humanoid dogs with antennas), "Land in New York or southern California so we won't look too conspicuous."

Apart from its regional earthbound jibe, the comic strip repeats once more the implicit assumption made by countless cartoonists, science fiction writers and movie producers that aliens are indeed in the Earth's vicinity considering how to make contact with us dumb Earthlings. Judging by the box-office receipts of films ranging from *Independence Day* to *Close Encounters of the Third Kind*, the idea has enormous popular appeal. But scientific opinion has been generally negative, with few outright advocates of the they-are-nearby-and-watching-us school.

I believe the objections to this popular notion are based more on emotional than scientific grounds, because recent findings in such diverse fields as astrophysics, biology, computer technology, planetary geology and spacecraft-propulsion technology all support it in various ways. I also feel that one of the least plausible scenarios is that of creatures' signaling us with radio transmitters from a planet of a nearby star. It seems to me far more likely that such random contact is the last thing alien intelligences would want to achieve. But before exploring these points in any more detail, I need to preface my remarks with some of the concepts that form the foundation of current scientific discussion of extraterrestrial life.

The most compelling support for the idea that we are not alone is the overwhelming vastness of the cosmos. Our own galaxy, the Milky Way, contains 50 times as many stars as there are humans on Earth. And there are billions of similar galaxies in the known universe. The total stellar cargo in the universe could easily exceed 10 billion trillion stars. Modern theories of star formation predict that at least some stars should have planets, and in the late 1990s, a few other solar systems were finally uncovered. But we are a long way from being able to estimate what fraction of those planets would provide the necessary conditions for the emergence of life. And we are even further from knowing what percentage of life-bearing planets could harbor intelligent life.

Yet as small as the percentages might be, we are still dealing with the residual from 10 billion trillion initial possibilities. True, nature is known to be profligate in these matters. Some species of plants disgorge billions of seeds in the likelihood that only one will grow to maturity. Could our universe be a garden with 10 billion trillion seeds and only one mature plant? I side with those who suggest that intelligence, having evolved once, has occurred elsewhere.

Just to pursue this line of thinking a bit further, suppose that during the history of the Milky Way Galaxy, at least a few other intelligent technological civilizations arose on places much like Earth. What might these civilizations be like? Because the majority of stars in the Milky Way Galaxy are several billion years older than the Sun, most alien life, on average, should be more evolved than we are. (Earth and the Sun formed virtually simultaneously 4.6 billion years ago; the oldest stars may have existed for more than 13 billion years.) We may be the new kids on the block. Unless all higher forms of life are ultimately

In a distant sector of the Milky Way Galaxy, a planet about the same size as Earth but only one-tenth its age has the potential to become life-supporting. Impacting comets and erupting volcanoes are already supplying water and atmospheric gases. If water, the crucial ingredient, remains liquid over the next few billion years, intelligent creatures may arise and begin their explorations of the galaxy.

Low-mass and high-mass extremes among Earthlike planets are suggested in these two renderings. The low-gravity Marslike world, left, barely clings to a life-sustaining atmosphere. Conversely, the high-gravity watery planet, right, has oceans so deep that barges of biomass remain as the only "land." On Earth, ocean creatures have existed for far longer than has land life, yet technological development is the preserve of land-based life. If such limitations are universal, technologically advanced life can exist only on Earthlike worlds with combinations of liquid water and solid land.

self-destructive, there has been plenty of time for alien civilizations to develop far beyond our level.

Self-destruction is a disturbing but plausible end point for technologically upwardly mobile intelligences. Just as a supernova is nature's way of terminating a massive star, self-annihilation may be a natural end to the evolution of life. However, I prefer to think that evolution is not inherently a dead-end affair and that *some* civilizations survive the bottleneck.

Other Earths

Of course, for a civilization to arise at all, we assume that it requires an Earthlike planetary environment. This assumption may be a failure of imagination on our part, but it is the only real starting point we have. And even at that, astronomers couldn't say for sure until the 1990s whether *any* other planets existed beyond our solar system. Compared with a star, a planet is like a firefly next to a searchlight. Shining by reflecting light from its parent star, a planet is simply too dim to be detected by direct telescopic viewing. Instead, astronomers analyze the target star's spectrum for telltale shifts that are caused by the orbiting planet's gravity pulling the star alternately toward and away from Earth. This technique is not sensitive enough to detect Earth-sized planets—yet. So far, only objects about half Saturn's mass or larger have been uncovered. The first success came in 1995, and search teams have compiled an impressive list of Jupiter-class planets orbiting other suns (see table on page 173).

Astronomers had long suspected that other planets must be out there and had assumed—correctly, it turns out—that their discovery would simply require a powerful enough telescope. The planets discovered as of 2004 range from about half the mass of Saturn to 10 times Jupiter's mass, which was pretty much as expected. But when astronomers calculated the planets' orbits around their respective stars, the results were anything but expected.

Instead of finding the Jovian-class planets in orbits like Jupiter and Saturn—the two largest planets in the solar system—the planet hunters were stunned to find that almost all the new planets orbit closer in to their parent stars, most of them *much* closer. One-third of the 122 planets discovered through mid-2004 circle their stars at less than the distance of Mercury, the closest planet to the Sun. Most of the others range from Mercury's distance to roughly three times the Earth's distance from the Sun. Moreover, many of these newfound worlds trek around their stars in highly elliptical (oval) orbits, unlike the nearly circular orbits of the planets in our solar system.

How did these planets assume such tight orbits? Although several scenarios have been proposed, theorists agree on one thing: Giant planets can't form at such close distances to a Sunlike star. Light gases, particularly hydrogen and helium, which make up the bulk of giants like Jupiter, would have been blown out of the inner region of an emerging solar system by the star's solar wind long before a Jovian-class planet

Our concept of a solar system—one star and a string of worlds orbiting it—is probably just one of a vast range of sun/planet combinations. Alpha Centauri, our nearest stellar neighbor, is a twin-sun system with yellow and orange components. In the illustration at left, the dual suns produce a double glow at sunset on a hypothetical Earthlike planet. Alpha Centauri has not yet been ruled out as a system capable of harboring some form of life. Several of the newfound super-Jovian planets, illustration above, with up to 10 times Jupiter's mass, could easily have moons big enough to retain extensive atmospheres—and possibly life.

could bulk up. Instead, theorists propose, the newfound planets formed in the same zone where giants are located in our solar system. From there, they either migrated inward or were thrown in.

Migration could be caused by a high-density primordial nebula that impedes the planet's orbital revolution around the star, forcing it to descend into a smaller and smaller orbit. Alternatively, several giant planets with roughly the same mass might form initially, but since no single planet gravitationally dominates, the system is inherently unstable over long time spans. Eventually, two or more of the giants will have a close encounter, which can send the planets into the inner region close to the star or eject them completely into endless treks between the stars.

Calculations show that a planet like Earth—one at a distance from its star that would permit liquid water—could not exist around any of the stars that harbor these newfound planets. The gravitational influence of a big planet in the inner region around the star would not permit a stable orbit for an Earthlike planet. Similarly, a giant planet farther out but in an elliptical orbit acts as a gravitational bully, upsetting the neighborhood and mauling its companions as it repeatedly swings closer, then farther from its sun. A smaller planet like Earth would not remain in a stable orbit for long and could be tossed completely out of its orbit and off into deep space.

The technique used to search for planets of other stars is biased toward detection of big planets close by the target star, so we have yet to see a true solar system analogue with giant planets in circular orbits at distances comparable to Jupiter and Saturn. But at least we can say that other planets do exist and that many of them orbit stars similar to the Sun.

Other investigations have uncovered what are almost certainly solar systems forming—rings of dust and gas surrounding newborn stars, just what would be expected from planet-formation theory. The key unanswered question now seems to be not *if* planets form but how rare Earthlike worlds are.

Interstellar Travel

That we are exploring other planets and moons in our own solar system places us at a pivotal evolutionary juncture which sentient creatures on other worlds either passed long ago or have not yet reached. The technological advancement from steam engine to computer over the past two centuries took but a metaphorical fraction of a second in the life of the universe. The swiftness of technological progress that we witness during a single human lifetime means the odds must be heavily stacked against two civilizations in a galaxy being anywhere close to the same technological century. As the late Carl Sagan once said, "To us, they would be either gods or brutes." Societies that evolved before us would regard the discovery of interstellar travel as ancient history.

Given time and the appropriate technology, there is no reason extraterrestrials could not travel around

the galaxy or beyond. Lengthy interstellar voyages could be achieved by retarding the biological clock that controls the aging process, by making the ship large enough to accommodate a microcosm of civilization, by sending surrogates in the form of robots or by avoiding the time factor altogether by traveling very close to the speed of light to utilize the time-dilation effect. Fusion or matter-antimatter propulsion systems could tap virtually limitless energy sources, permitting unbounded exploration.

But after space travel becomes commonplace and problems of health and energy abundance are solved, one question would remain for any curious civilization: Does the universe harbor other rational creatures? Unless advanced civilizations are spread so thinly that there are fewer than one per galaxy, they will become aware of each other sooner or later. Either individually or collectively, the civilizations which arose before us in our galaxy must have thoroughly surveyed their interstellar environment, if for no other reason than to attempt to answer that question. Long ago, the process would have led to the discovery of life on Earth.

Despite the fact that our planet is a relative newcomer on the galactic timescale, the oceans have nurtured life-forms for three billion years, about one-fifth of the age of the galaxy—ample time for Earth to have been noticed as a life-bearing world. This may sound like the background for a science fiction movie, but consider the alternatives: If aliens are not aware of us, they must have all self-destructed or they are not interested or we have always been alone or they are there but have not yet discovered us. Let's consider each of these possibilities in turn.

The weakness in the self-destruct scenario is that it is too restrictive. To suggest that no civilizations would survive is to assume that self-destruction is a fate sealed in the cells of primordial organisms on every life-bearing world—a closed loop, like the birth-and-death cycles of massive stars. Similarly, the idea that all alien intelligences are uninterested in seeking other life is too narrow an assumption. It ignores us, the only

Hundreds of billions of stars populate the Andromeda Galaxy, the nearest major galaxy to the Milky Way. The universe contains at least 100 billion galaxies, with a total cargo of roughly 10 billion trillion stars. Unless modern theories of star formation prove to be wrong, many of those stars should have planets. During the 1990s, astronomers gathered the first solid evidence for Jupiter-mass planets orbiting Sunlike stars within our galaxy. Now the search is on for Earthlike worlds.

example of technologically advanced life we know of.

The third possibility, that we have always been alone, is also unlikely, given the billion trillion stars in the known universe. The origin of life would have to have been a once-only affair, which would put us in a unique position in the universe, at the apex of cosmic creation. There is some precedent to suggest that this may be an unwise viewpoint. All previous theories that placed humans in a central or preferred position in the celestial hierarchy have proved false. There was the pre-Copernican Earth-centered universe, the post-Copernican Sun-centered universe and, early in the 20th century, inflated estimates of our galaxy's size and special location.

A variation on the we-are-alone theme is that a steady supply of potential advanced life is always emerging in the universe but never survives due to natural but externally induced catastrophes, such as severe meteorite bombardment, nearby exploding stars or passage through a dense nebula. In fact, several devastating waves of extinctions on Earth obliterated many life-forms and markedly changed the course of evolution. Maybe we have just been lucky—a jackpot on the cosmic slot machine. An intriguing idea, but it relies on chance events producing a unique exception, clearly an unwise assumption.

The final alternative assumes that no other civilizations explore beyond their local neighborhood. Supporters of this view argue that interstellar travel is technologically difficult and enormously expensive and that consequently, it would not be commonly done. I have always thought that such arguments express "difficulty" and "expense" in terms of 20th-century or, at best, 21st-century technologies. Even the most perceptive visionaries, such as H. G. Wells and Leonardo da Vinci, projected technological progress only a few centuries. No one alive today can possibly guess what devices will propel us—or our consciousness—to the stars. In principle, interstellar travel does not defy the laws of physics. Or as noted science fiction author Arthur C. Clarke has said: "Any sufficiently advanced technology will seem to us like magic."

The Quest for Signals

Interstellar travel is the most direct way to seek life on other worlds. Nothing can equal actually taking a close look. And firsthand investigation is probably the only way that creatures less technologically developed than humans could be detected. Searching for alien signals with radio telescopes will yield results only if other civilizations are staying at home making long-distance calls to their galactic relatives via radio frequencies. Radio-search advocates not only admit this but have enshrined the concept in what Princeton University physicist Freeman Dyson calls "a philosophical discourse dogma." He says the radio searchers assume "as an article of faith" that higher civilizations communicate by radio in preference to all other options.

Prior to 1959, there was no dogma on the search for intelligent extraterrestrials. All references to the subject were found only between the covers of science fiction books and magazines, where interstellar spaceships were an integral part of the picture. Then, almost overnight, everything changed. It became scientifically respectable (or, in the view of some, tolerable) to talk about contacting aliens because the prestigious journal *Nature* published an article by physicists Philip Morrison and Giuseppe Cocconi describing the relative ease of transmitting radio signals across the galaxy and how it might be possible, using radio telescopes, to detect such signals from other intelligences. Hundreds of articles and dozens of books embellishing the concept have appeared since this watershed article, but the basic idea has remained constant.

The first actual search, by astronomer Frank Drake, was conducted a few months after the Morrison-Cocconi article was published. Along with Cornell University's Carl Sagan, Drake went on to become the subject's most visible spokesman. Recognizing that interstellar spaceships and radio searches are concepts in conflict, Drake advanced the often-repeated argument of the extreme difficulties and the enormous expense involved in interstellar travel. But that was in the early 1960s. Since then, dozens of serious research reports and proposals have detailed various interstellar propulsion systems that are considered reasonable extrapolations of late-20th-century technology. That weakened the rationale for the radio searches, and by the mid-1970s, a significant number of astronomers were calling for a reassessment of the assumptions. In the ensuing discussions in scientific journals, those interested in the question split

into two camps: the agnostics and the dogmatists. Scientists from one group are seldom invited to SETI (search for extraterrestrial intelligence) conferences organized by the other.

Alien Visions

Radio-search advocates have one powerful comeback: If we do not listen, we will never know for sure. I do not object to someone's monitoring the cosmic radio dial; however, the public's perception of the activity is that researchers are attempting to eavesdrop on alien conversations, which is not what the radio quest is all about. Even if the galaxy were humming with alien radio communications, we almost certainly would not intercept any of them unless the transmissions were outrageously extravagant in signal power, which goes against the initial argument about the efficiency of radio communication. Signals directed from point A to point B in the galaxy would be undetectable unless we happened to be precisely between the two points—an enormous improbability. Our radio telescopes are far too weak to pick up conversations not focused in our direction.

Could we recognize other kinds of signals? Earthlings have broadcast their existence for about 70 years through radio, television, radar and other electronic transmissions that escape into space. This "noise" floods out from our planet at the speed of light and now form an expanding bubble of babble 70 light-years in radius. But could we detect the noise from another civilization? According to a 1978 study conducted by Woodruff T. Sullivan and his colleagues at the University of Washington, existing equipment would be able to pick up electronic "leakage" of television- and radar-type emissions from only a few dozen light-years away—a radius that includes fewer than 200 stars. In the decades since that study, the sensitivity of our receivers has advanced to extend the detection radius to include thousands of stars, but this affords us little advantage, because electronic leakage from an Earthlike planet would be swamped within a few hundred light-years by the static of natural cosmic radio sources.

In any case, such leakage would not last for long, according to Sagan. "I think there is just a 100-year spike in radio emissions before a planet becomes radio-quiet again," he said in a published interview in the 1980s. He pointed out that communications on Earth are rapidly moving from brute-force broadcasting to narrow-beam transmission, satellite dishes, fiber optics and cable television—all of which are

Two marginal environments for life: a planet of a close double star, above left, and the atmosphere of a gas-giant planet, above right. Close double stars often affect each other's evolution through matter transfer, as in the system RW Persei, illustrated here. Double stars with masses similar to the mass of the Sun and somewhat farther apart than RW Persei could, in theory, have an Earthlike planet with a stable orbit. But whether the orbit would remain stable for billions of years is questionable. Facing page: A radio telescope scans the distant stars.

Greek scholars of the third century B.C. were aware of the Earth's spherical shape, but to them, a view such as this Galileo spacecraft image of Earth and the Moon seen from afar could be regarded only as magic. Similarly, a technology two millennia in advance of us could have developed methods of remote surveillance we cannot begin to imagine.

relatively low power. Sagan's assessment: "The chance of success in eavesdropping is negligible."

Despite the frequent use of the word "eavesdropping" in connection with the SETI programs, most of the searches have been limited almost entirely to finding either a superpowerful omnidirectional beacon or a narrow-beam signal intentionally directed toward us. Such a search strategy therefore assumes that extraterrestrials either know about us or are pumping a colossal signal in all directions in the hope that somebody is listening.

A number of thorny questions emerge from all of this. How long would a civilization continue to broadcast an omnidirectional signal? Since it would take hundreds of thousands of years for a response, would everybody be listening and no one sending? And why would an alien civilization want to broadcast, not knowing who would intercept the message, what culture shock it would cause or who might be around on their home planet to receive a reply centuries or millennia in the future? If a signal is intentionally beamed in our direction, the senders must know we are here. Is the universe so perverse in its structure that it constricts intelligences to a destiny of exploration by megaphone? Is it possible that extraterrestrials might consider radio transmission to be as quaint as we regard smoke signals or jungle drums? Throughout human history, many methods of communication have been relegated to the technological scrap heap.

Nevertheless, approximately 50 radio-telescope searches have been undertaken since Drake's first attempt in 1960. Since 1984, a refurbished radio telescope in Massachusetts has been scanning for artificial interstellar signals 24 hours a day, a project privately funded by The Planetary Society, a California-based organization of enthusiasts interested in continued exploration of the planets and beyond. University of Maryland astronomer Ben Zuckerman, who conducted the largest survey prior to The Planetary Society's effort, now thinks the chances of such searches succeeding are so remote that he is no longer pursuing the activity. The late Iosif Shklovskii, an eloquent search enthusiast who coauthored the classic *Intelligent Life in the Universe* with Carl Sagan in 1966, also grew disenchanted a decade after the book was published.

In 1978, at the height of scientific interest in the radio searches, Sagan wrote his most optimistic article on the quest for an intentional signal from another civilization. "Since the transmission is likely to be from a civilization far in advance of our own," he wrote in *Reader's Digest*, "stunning insights are possible, [including] prescriptions for the avoidance of technological disaster. Perhaps it will describe which pathways of cultural evolution are likely to lead to the stability and longevity of intelligent species. Or perhaps there are straightforward solutions, still undiscovered on Earth, to problems of food shortage, population growth, energy supplies, dwindling resources, pollution and war."

And Sagan was not the only one to express such a view. Frankly, I think suggestions that messages from aliens will solve all our problems not only are presumptuous but also smack more of religion than science. Surely superior intelligences in the universe have graduated to something more creative than operating broadcasting stations to send galactic versions of the Ten Commandments to the heathens.

Our own experience on this planet suggests that contact between technologically imbalanced cultures is bad news for the number-two culture. Painful assimilation and cultural decimation seem to be the products of such contact, even with the best of intentions. Maybe there is some biogalactic law among advanced civilizations that forbids direct intervention

with primitives like us. In any event, technologically superior intelligences could probably make themselves completely undetectable, leisurely learning all they care to know about life on Earth without our being aware of the scrutiny.

Their physical appearance may also be unrecognizable to us. They may once have passed through an evolutionary phase when they resembled humans, but the lizard men and the gaggle of other humanoids from the *Star Wars* and *Star Trek* movies are probably not who (or what) we will meet. Evolution beyond humanlike form is as inevitable as our ascent from our reptilian ancestors. One billion years ago, the highest form of life on Earth was the worm. An alien intelligence one billion years ahead of us on the evolutionary ladder could be as different from us as we are from worms.

On Earth, the next stage beyond humans will probably be computer, not biological, evolution. In some ways, computers can be regarded as a newly emerging form of life, one built on silicon rather than carbon (the basis for all biological life as we know it). A silicon computer "brain" can have unlimited size and capacity, whereas the human brain has not increased in size for at least 75,000 years. Continuing miniaturization of computer components could lead to a synthesis between humans and nonbiological computer intelligences. For example, a microcomputer might be surgically implanted in the human brain. One would merely think a question in a manner that the computer could understand, and the answer would be provided as a conscious thought.

Theoretically, there is no limit to how far this technology could progress. Perhaps bodies of bone and flesh are already redundant in the universe, and advanced civilizations have become virtually indestructible semi-immortal arrays of silicon or its evolutionary successor. To such a form of intelligence, time would have a totally different meaning. With no finite life span to impede time-consuming activities such as interstellar travel, millions of years could be spent in exploration. Voyages to countless star systems would present no problems for a semi-immortal brain. To such travelers, emerging intelligences like ours would be fascinating biological crucibles. Occasionally, they might look in on Earth to glimpse the latest tribal squabble and wonder when we will emerge to seek our place in the galactic community.

What About UFOs?

If aliens can travel here, then where are they? Where are their artifacts, their spaceships, their supply bases? So far, there is not a shred of hard evidence that extraterrestrials exist, either in our vicinity or in deep space. But absence of evidence is not evidence of absence. A century ago, it would have been impossible to identify a modern reconnaissance satellite—able to spot a human shadow in a playing field from an altitude of 200 kilometers—as anything more than a mysterious moving dot of light tracking across the night sky. Even that could be concealed by coating the spacecraft in ultralow-reflective material. Because there could be millions of years of technical advancement between Earthlings and aliens, it might be a simple matter for them to remain undetected. Yet thousands of reports of strange aerial phenomena—unidentified flying objects, or UFOs—have been cited as proof that extraterrestrial devices are exploring Earth.

I delved seriously into the subject after I saw a UFO in 1973 that was witnessed by 18 other people. I was teaching a class in astronomy at a planetarium in Rochester, New York. We were outside identifying constellations when we saw a formation of lights pass silently almost overhead, then veer off and swoop toward the horizon. It is not impossible for an aircraft

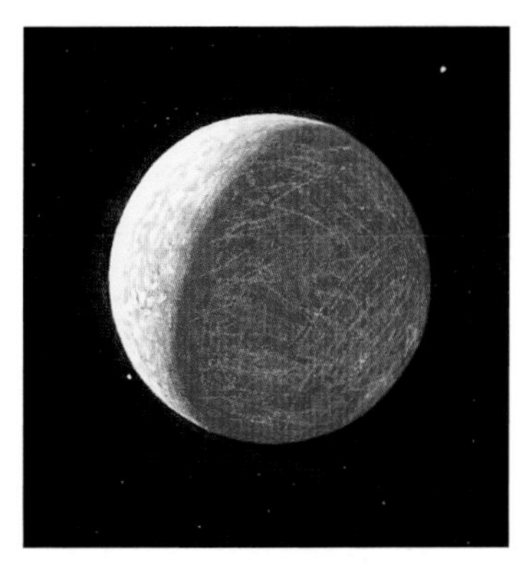

Left: A planet several times the mass of Jupiter orbiting a red-dwarf star could be the most common type of planetary system in the universe, because red dwarfs outnumber Sunlike stars by more than 10 to 1. Using indirect detection techniques, astronomers have already found about two dozen Jupiter-sized and larger planets. Some, such as the one rendered above, are just a fraction of Mercury's distance from their parent stars—close enough for their atmospheres to glow in the intense heat.

Above: Alpha Centauri, the nearest star to the Sun, at a distance of 4.4 light-years, appears crammed among throngs of fainter stars. Yet this is a complete illusion—the immense distances (up to 10,000 light-years) separating the individual suns are not obvious in a flat photograph with no depth cues. The average distance between stars in our sector of the Milky Way Galaxy is seven light-years. Adrift in this vast ocean is Earth, facing page, which would clearly announce its inhabitants to its neighborhood. In addition to the city lights, a flood of electromagnetic signals (radio, radar, television) issues from our planet every second. Apart from that, the oxygen in our atmosphere has been a telltale sign of life for more than a billion years. If aliens exist, they almost certainly know about our life-bearing globe by now.

or a group of aircraft to behave in this manner, but I had never seen that kind of formation nor the type of lights we observed.

What made the sighting even more baffling was being told when I called the air force base responsible for the local airspace that no military vehicles were anywhere near the area at the time in question. A controller at the local airport said the same thing. Only through persistent sleuthing did I finally discover that what we had seen was, in fact, a small fleet of experimental military aircraft with high-intensity downfacing lights, apparently top secret at the time.

If I had been unable to take the time and effort to research the case properly, I would have been left with a lifelong UFO story to tell. That is not to say every UFO which has ever been seen can be explained in a conventional manner, as some debunkers would have us believe. A small number of cases have been thoroughly investigated and remain a mystery. These may represent a hitherto undiscovered phenomenon, perhaps a rare atmospheric electromagnetic effect. Extraterrestrial spaceships capture the imagination because they are the most dramatic of the possibilities.

Some accounts of UFO and flying-saucer sightings have become modern legends. They include explicit descriptions of alien creatures and, in a few cases, bizarre tales of aliens that allegedly abducted humans for hours or days. I became sufficiently intrigued by such reports that I interviewed some of the "abductees." Despite the sincerity of these people, I remain unconvinced—and I was ready to be convinced. I have no bias against the ability of extraterrestrials to traverse the gulfs between the stars to explore Earth, but there are too many inconsistencies. Why, for instance, would they allow themselves to be seen as UFOs—elusive, yet hinting at so much—playing peekaboo with the natives? Are we to conclude that aliens choosing to venture close to Earth are *almost* clever enough to go undetected, but not quite? I think not.

Over the years, I have investigated many UFO sightings and was, for a time, a consultant for a major scientific UFO study, which forced me to give serious thought to the matter. I have concluded that the investigation of UFOs is not a fruitful avenue to pursue in the hopes of contacting our cosmic cousins, although it may prove of value in some other way.

In the late 1960s and early 1970s, discussions of UFOs entered the pages of scientific journals and engaged the interest of a cross section of researchers ranging from astronomers to psychologists. But that era seems to have ended, and UFOs have virtually disappeared from the pages of scientific publications.

Contact: Alien Communication

The standard science fiction stories of aliens invading Earth represent, in my view, the least likely scenario. Cultures more advanced than ours will be benign; they will have learned to live with themselves. Aggressive creatures prone to squabbling and warfare would be weeded out by natural selection—they would destroy themselves. If by some fluke they did not, they would be kept in check by even more advanced civilizations acting as galactic police. Otherwise, we would have known about them by now, because creatures with the will and the power to enslave or destroy us would have done so long ago.

We may be a celestial nature sanctuary that has reached a critical evolutionary stage. For the first time, the specter of self-annihilation looms as a possibility. We would be of prime interest to alien intelligences. To understand why, consider the opposite point of view. When we have the ability to travel to solar systems beyond our own, the most exquisite discovery we could make would be other living creatures. Even recognizing a lowly bacterium on another world will be a momentous event, telling us at last that we are not alone. But that revelation would pale in comparison with the detection of a world harboring creatures which might eventually evolve to contemplate their own existence. In any conceivable circumstance, thoughts of conquest or destruction of such a crucible would be totally pointless.

The most valuable thing we Earthlings have to offer advanced aliens is ourselves in our natural state. We are a biological and scientific oasis, a living example of intelligent beings climbing the evolutionary ladder. We are at the crucial stage where we will either become spacefaring creatures or plunge ourselves into a catastrophe of our own making. Whatever the outcome, I believe our cosmic relatives are likely aware of us and are observing our progress with interest.

Passive observation and nonintervention are the only approaches that would pay reasonable dividends for extraterrestrials. Despite efforts to eavesdrop on extraterrestrials with radio telescopes, the odds favor the belief that the aliens already know about us and are silent observers. We will remain unaware of them until they are ready to talk. Contact will be made at a time and in a manner of their choosing, not ours.

HOW THE UNIVERSE WILL END

CHAPTER NINE

This is the way the world ends. Not with a bang but with a whimper.

T.S. Eliot

Questions of origins and destinies have always held a deep fascination for humans, as evidenced by the principal preoccupation of virtually every prophet and religious doctrine. Cosmology deals with the same concerns on a cosmic rather than a human scale, although theological overtones are not entirely absent.

The Big Bang creation of our universe set our cosmic clock in motion some 14 billion years ago. At that instant, time as we know it began. Understanding what happened since then is the essence of modern astronomy and cosmology. Because we can probe the past by peering to the far reaches of the cosmos, the history of our universe is written in what we see at ever-increasing distances. But the universe's future has yet to be. It can be neither seen nor measured, only inferred from what has already happened.

Everybody likes a detective story, and there are solid clues in this one that allow some informed guesses about our ultimate cosmic destiny. The starting point for cosmologists was agreeing on the fundamentals of the universe's origin. That came in the late 1960s, when many astronomers were persuaded that the Big Bang scenario was the best explanation for the expanding universe we see around us. Two obvious fates emerged as contenders: The expansion will continue forever (known as an open universe); or the expansion will eventually be halted by the universe's overall mass, and a collapse will ensue, culminating in the Big Crunch (a closed universe).

In the 1980s, inflationary theories predicted a third alternative: a flat universe, in which the outrush will slow to an equilibrium, with the final destiny being neither endless expansion nor collapse.

In recent years, the pendulum has begun swinging more in the direction of the open universe. But to keep ourselves in neutral territory for a moment, let's flesh out the skeleton ideas of the three scenarios of the universe's ultimate destiny.

A closed universe is clear-cut. A few tens of billions of years after the Sun becomes a white dwarf, the expansion will cease. Like a movie running in reverse, the galaxies will begin to approach one another, first gently, then with escalating violence as they merge into a fierce fireball of supernovas and superquasars. Nothing can prevent total collapse in such a conflagration, and the entire affair will swallow itself in a black hole the mass of the universe. All that we know today will be compressed to a singularity smaller than the nucleus of an atom. Time and space will no longer exist.

Some cosmologists have sidestepped the finality of the closed universe by suggesting that the collapse of our universe will provide energy for the creation of another. Our universe then becomes one bead in an infinitely long string of birth, death and rebirth. The oscillating-universe hypothesis has great appeal because it embodies a dynamic, evolving universe with a definite life span, yet it never dies. Instead, it reincarnates, phoenixlike, from its own ashes. But in recent years, certain theoretical difficulties have clouded the oscillating-universe idea. Despite efforts by numerous cosmologists, no one has offered a convincing reason why the collapse will not be the final curtain for our universe.

Less tidy than the life cycle of the closed universe is the final blackness that awaits both the flat and open scenarios. Neither has a definable end; the universe just fades away. Galaxies will continue to man-

Like a colossal flower hanging in the fabric of space, the Orion Nebula, left, blooms on the edge of a vast cloud of gas and dust that extends far beyond the area pictured. The nebula's illumination stems from four massive stars in a compact group called the Trapezium, just left of center. Nebulas like Orion are constantly renewing the cycle of stellar birth and death within galaxies. But it will not go on forever. The most likely fate of the universe is a slow fade to eternal darkness.

Facing page: Ablaze with distant suns, the dome of the night sky is bisected by the powdering of the Milky Way in this ultrawide-angle photograph taken from the summit of Mount Graham in Arizona. The North America Nebula, just above center, is shown in the close-up above. Every one of the stars in these two views would have been visible when the Great Pyramid of Khufu was being constructed. Their positions and brightnesses at that time were but slightly altered from what we see today. In comparison to human life spans, the stars seem as enduring as mountains. But we now know that over vast stretches of time, neither stars nor mountains are permanent.

ufacture stars from the dust and gas within them for at least another 20 to 30 billion years. But after that, when the nebula gas is depleted and new stars cease to be created, the only stars that will continue to shine are the least massive red dwarfs, like Proxima Centauri. If Earth escapes being vaporized during the Sun's red-giant phase, the sky of this distant future will be as black as the inside of a cave, at least to the naked eye. In that long night of the future, perhaps 40 billion years from now, the Sun will cool from a white dwarf to a dark, dense lump—a black dwarf—undetectable except for its gravity, which will continue to hold Earth in its orbit.

Red-dwarf stars with masses from one-quarter to one-tenth that of the Sun burn long and slow, some lasting for more than 10 trillion years before finally sputtering out. But by 100 trillion years from now, the last stars will have winked out. The universe, containing the darkened and collapsed corpses of a trillion trillion once brilliant suns, will finally fade to black.

Almost all matter in this distant future universe will be locked in stellar tombstones: black holes, neutron stars and black dwarfs. The remainder will be in the form of planets, asteroids and comets. If current predictions in subatomic physics are correct, most of the universe's protons will have decayed by 100 trillion trillion trillion years from now, causing everything except black holes to decompose. Only the black holes will remain, leaving the universe as a desolate void punctuated with invisible gravity whirlpools.

On the timescale of the universe as we know it, where ages are measured in billions rather than trillions of years, black holes are immutable. But given enormous spans of time, they, too, will decay through a process called quantum tunneling. A solar-mass black hole would take a million billion trillion trillion trillion trillion years to snuff itself out. More massive black holes would take longer, but eventually, the universe would become an exceedingly tenuous mist of subatomic particles and radiation far closer to a vacuum than intergalactic space is today. In this bleak future, the temperature everywhere would be less than one-trillionth of a degree above absolute zero.

One possible escape from this long, slow fade to black is the recently revived cosmological constant discussed in Chapter 7. If it exists as a form of repulsive dark energy hidden in the vacuum of space, as new observations suggest, this throws an additional factor into the scenario. For instance, if the cosmological constant is another form of a phase transition similar to the one that occurred in the Big Bang's inflationary era, all bets are off. Rather than going out with a whimper, the universe could ultimately transform itself into something utterly different—as different as our universe is from what came before the Big Bang.

Contemplating the Cosmos

Cosmology can say nothing about the fate of life over the universe's history. Isaac Asimov's provocative 1956 science fiction short story "The Last Question" (which he described as his finest work) offers an intriguing prospect. In the remote future, long after the last star has ceased to shine, all intelligences in the universe have combined into one omniintelligence that has no physical form but pervades the very interstices of space. Seeing the final black death of the universe everywhere, the intelligence decides that there is no alternative but to rework it into a new universe. The preparations complete, the plan is fulfilled as a new Big Bang erupts from the vacuum. At the height of his creative genius, Asimov was 36 when he wrote this story, and it has always impressed me.

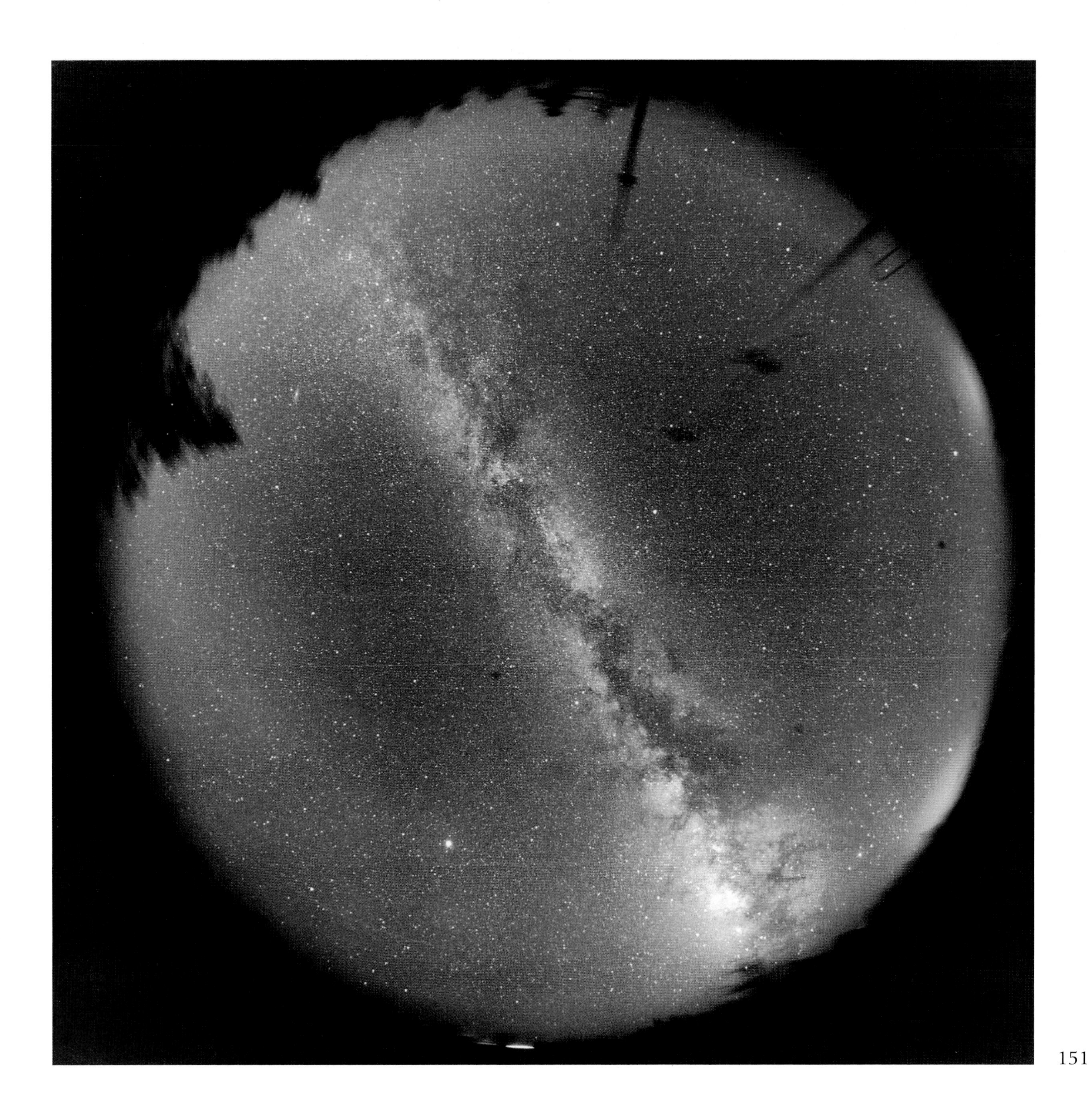

The universe knows nothing of our notions of a creator or of our quest for order and our distaste for a cosmos that fades into oblivion. Or does it? Some cosmologists have suggested that the universe exists only because we are here to perceive it—a concept dubbed the anthropic principle by its proponents. They assert that universes where the fundamental constants of nature do not allow stars, planets and life to emerge cannot exist. This notion is dismissed by most astronomers; Carl Sagan called it "bizarre." Yet it has its adherents. A few years ago, I attended a three-day international conference where these ideas were discussed and debated by astronomers, anthropologists, philosophers and theologians. I confess, I came away not much wiser.

A few days later, on a tranquil summer evening, I spent several hours in the backyard with my favorite telescope. The sky was wonderfully black. Rivers of stars in the Milky Way paraded across the telescope's field of view. Saturn's rings were sharp and clear and as compelling as ever. In the Virgo galaxy cluster, nine galaxies appeared in one field, six in another. I found myself returning to the questions debated during the conference. There is no escaping it: We *need* to know how we fit into the big picture. Now that we have a reasonably accurate idea of the extent and content of the universe, our significance in it becomes the ultimate quest that will drive our descendants to explore those distant stars.

Perhaps our current cosmological concepts will seem as quaint to astronomers of the 22nd century as those of the 18th century seem to us. Two hundred years ago, the universe of galaxies was totally unknown. Our own galaxy's shape and dimensions could only be guessed at. Two hundred years from now, all aspects of the universe's evolution from birth to death will be mapped in detail. Although much of what we know today will become the foundation blocks for the more far-reaching cosmologies of the future, some of today's theories are destined to become historical footnotes. Nobody wants to devote a career to a long-term loser. But it has happened many times. What is "hot" now may or may not be relevant in a decade.

One of the alluring aspects of modern astronomy is that several significant revolutions in thought as well as many new discoveries can occur in a single lifetime. But that is a recent development. From the dawn of civilization until about 20 generations ago, Earth was regarded as the focus of creation, the Sun, Moon and stars as magical phantom lights on the crystalline dome of heaven, somewhere out of reach. To our forebears, the universe was little more than a giant planetarium, its machinery mysterious, its meaning shrouded in myth and superstition. For centuries, nothing happened to alter the situation. Upstarts who challenged the status quo were quickly silenced.

Facing page: In a timeless cycle, Earth spins on its axis, twirling the universe before us. The dome in the foreground is at Las Campanas Observatory in Chile's Atacama Desert. Left: One of the estimated 100 billion galaxies in the known universe. Like snowflakes, each spiral galaxy has a unique shape and structure, offering almost infinite variations on a single elegant theme.

Today, the reverse is true. For the first time in human history, it has become impossible for a citizen of planet Earth to digest the whole scope of astronomical knowledge. Trying to keep up with current discoveries is like swimming upstream: With great effort, one can do it for a short time, but inevitably, it overwhelms the best attempts. And the questions flow as rapidly as the answers. Maybe it will always be so. If not, it means we live in a very special time.

One thing can be said for certain: The past few decades have been the most remarkable in the history of astronomy. Through space-probe investigations, we have come to know the planets in a way unimagined two generations ago. Cosmology has graduated from the arm-waving speculation of the 1920s to a fairly rigorous science with a firm foundation of facts upon which theories can be built. We live in a golden age of astronomical discovery—perhaps *the* golden age. And the best part of it all is that we are here to experience it firsthand.

TELESCOPES FOR THE 21st CENTURY

It is unwise to trust theory very far when it becomes divorced from observational test.

Sir Arthur Eddington

Perched above 40 percent of the Earth's atmosphere on the barren summit of Hawaii's Mauna Kea, an extinct volcano towering 4,200 meters above sea level, astronomers using the largest collection of research telescopes on Earth are as close to astronomical nirvana as they can get without leaving the planet altogether. The optical arsenal includes the twin 10-meter Keck telescopes, the 8.3-meter Subaru reflector, the 8.1-meter Gemini telescope and a dozen smaller instruments, their protective white domes resembling impossibly huge dinosaur eggs.

Mauna Kea is one of two sites considered the best in the world for ground-based telescopes. The other is in and around the Atacama Desert in northern Chile, where five mountaintops bristle with observatory domes. On one of them, Cerro Paranal, four huge observatories each house an 8.2-meter telescope. Known as the Very Large Telescope (VLT), the giant instruments constitute the largest telescope array on Earth. By 2007, astronomers expect to have starlight collected by all four instruments relayed to arrive at a single focus, thereby combining the light-collecting power and creating, in effect, a single 16-meter telescope.

All the giant telescopes in Chile and Hawaii were built in the late 1990s as part of several breakthroughs in telescope design and construction that represent a major advance in astronomical firepower which is now taking us into the new century. The telescopes of the previous generation, epitomized by the famous 5-meter (200-inch) Hale Telescope on California's Palomar Mountain, were the workhorses of astronomy from the 1950s into the 1990s. Characterized by massive equatorial mounts designed to glide smoothly to compensate for the Earth's rotation, these elegant giants were the culmination of precomputer-era engineering. The modern generation of telescopes utilizes squat altazimuth mounts fully computerized to do the same precision tracking of celestial objects as do the mammoth equatorial mounts.

The second major revolution in the new telescopes—main mirrors up to 11 meters wide—also emerged from modern computer technology. To retain their precision shape, previous-generation telescope mirrors had to be thick enough not to bend under their own weight. At 16 tons and nearly a meter thick, the Hale Telescope's 5-meter mirror proved to be the practical limit.

One solution, applied to the twin Keck 10-meter telescopes on Mauna Kea and to the 11-meter Hobby-Eberly telescope in Texas, is to use dozens of precisely fashioned thin segments combined like tiles into one huge curved surface. Computer controls keep the segments perfectly aligned with one another.

Another approach—the one chosen for most of the new-generation instruments—employs one-piece mirrors up to 8.4 meters in diameter that are much thinner than the traditional mirrors. For example, the Subaru and Gemini telescope mirrors are about the same weight as the Hale 5-meter but are only as thick as a two-year stack of *National Geographic* magazines. They bend. But they also rest on a bed of 100 to 200 actuators—computer-controlled pistons that move in increments as small as 1/10,000 the thickness of a human hair. The actuators can make thousands of adjustments per second to ensure that the mirror retains its perfect shape for prime optical performance. This mirror-morphing technique is known as active optics.

Even so, the big telescope mirrors have a lot of air

The Hubble Space Telescope, the most powerful telescope ever built, floats in an orbit 600 kilometers above the Earth's surface. Although its 2.4-meter main mirror is small by today's standards, nothing on Earth can match the sharpness of the Hubble's images undistorted by the constant turbulence in the Earth's atmosphere.

About once every three years, a space shuttle crew visits the Hubble Space Telescope to replace obsolete or worn-out parts and to give the big telescope a checkup. Regular camera upgrades have kept the orbiting observatory operating at peak performance. The famous optical flaw that marred the mission after its launch in 1990 has been remedied, and the telescope is providing valuable scientific data beyond all original expectations. NASA plans to keep the Hubble active until at least 2007.

Perhaps the most dramatic example of the Hubble Space Telescope's capabilities is this pair of images of the Eagle Nebula taken by the venerable 5-meter Hale Telescope, above, and the Hubble, left. The Hubble shot is a close-up of the part of the nebula at center in the Hale shot. The color differences between the two images stem from the use of film emulsion (Hale) and CCD electronic imaging with filters (Hubble).

to look through as they peer into the cosmos, and this is potentially ruinous to their performance. The Earth's atmosphere is always turbulent to some degree, as evidenced by the twinkling of the stars seen on any clear night. Locating observatories at high altitude avoids the worst turbulence in the lower atmosphere, but enough remains to distort telescope images significantly. To combat this problem, astronomers use another computerized technique, called adaptive optics, which monitors a star or galaxy in the telescope's field of view with an optical sensor. A computer then analyzes exactly how the image is being distorted by the atmosphere and, at lightning speed (up to 1,000 times per second), calculates how to neutralize the distortion. This information is sent to a flexible "plastic" mirror, about the size of the palm of a man's hand, placed in the telescope's light path just in front of the main camera. The flexible

Telescope heaven atop Hawaii's Mauna Kea mountain. At upper right are the twin domes of the 10-meter Keck telescopes. Just to their left is the angular "dome" of the 8.3-meter Japanese national telescope Subaru ("Pleiades" in Japanese). The large dome just below center houses the 3.6-meter Canada-France-Hawaii Telescope (CFHT), the first big telescope erected on this prized 4,200-meter summit. To its left is the open dome of the 8.1-meter Gemini North telescope, under construction when this photograph was taken. Details of the Gemini telescope are shown on the facing page (top left and right). After being reflected off Gemini's huge main mirror, starlight is then reflected off the small mirror at the top of the telescope, which focuses it toward instruments beneath the hole at the center of the main mirror. Gemini North is the last large telescope to be built at this site, which is now considered "full." The photograph at right was taken at sunset looking toward the Gemini dome from the CFHT. At lower left on the facing page is an aerial view of the Very Large Telescope in Chile, an array of four 8.2-meter telescopes that, when completed and optically linked together in 2007, will effectively be the largest telescope on Earth.

mirror compensates for the distortions in the atmosphere, restoring the sharp image of the star or galaxy.

These image-correcting techniques work astonishingly well, but nothing can match eliminating the Earth's atmosphere entirely by placing the telescope in orbit. The sharpness and clarity achieved by the Hubble Space Telescope is still unequaled by any instrument on Earth, despite the Hubble's relatively puny 2.4-meter main mirror. Moreover, the Hubble functions in the true blackness of space. Surface telescopes, no matter how well situated on remote mountaintops, are affected by airglow, a constant low-level aurora at a 40-to-100-kilometer altitude that gives the atmosphere a weak brightness which subtly hazes

Above: In 2012, NASA plans to launch the Next Generation Space Telescope (NGST), which boasts a 6-meter segmented main mirror that will open like the petals of a flower once the telescope is in space. The design includes a 30-meter-wide shade to keep the telescope in shadow and at a constant cold temperature for optimum performance. One major difference between the NGST and the Hubble Space Telescope is that the NGST will orbit the Sun, rather than Earth, which will keep it away from sunlight reflected by our home planet. Right: An artist's concept of the proposed Overwhelmingly Large (OWL) telescope, a 100-meter-diameter behemoth that would require a cooperative effort by many governments.

long-exposure images. Computer processing can substantially reduce but not eliminate the effects of airglow. Orbiting telescopes will always be the ultimate solution to the astronomical annoyances of our planet's life-giving blanket of air.

What lies ahead? With the Hubble's sensational record of unrivaled astronomical discoveries in mind, engineers are at work developing the Next Generation Space Telescope (NGST). Since telescopes larger than the Hubble won't fit into the space shuttle cargo bay or any other launcher to be developed in the foreseeable future, the solution is some type of mirror-unfurling arrangement. To keep it well away from heat and reflections from Earth, the NGST (now known as the James Webb Space Telescope) will orbit the Sun, too far out to be serviced by the space shuttle. Whatever the final design, astronomers hope to have a 6-meter telescope in space not long after the Hubble is expected to be retired—around 2010-2012.

Meanwhile, encouraged by what has been learned from the current generation of ground-based telescopes—the twin Keck 10-meter instruments in Hawaii; the two 8.1-meter Gemini telescopes in Hawaii and in Chile; the Japanese Subaru 8.3-meter reflector in Hawaii; the four European 8.2-meter Very Large Telescopes in Chile; and several others—astronomers are looking ahead and thinking big. To get a feeling for how big, note in the illustration above the tiny humans and vehicles in front of the proposed 100-meter OWL telescope.

Somewhat less ambitious, though still mind-boggling, is the Thirty Meter Telescope that is being suggested by a consortium of U.S. and Canadian universities. This segmented-mirror design is a Keck telescope scaled up to nine times the size (mirror area).

And the feasibility of a 50-meter telescope for construction on La Palma, one of the Canary Islands off the northwest coast of Africa, is being studied by a group of astronomers from five European countries. Several large telescopes are already located on La Palma, and researchers rank it as one of the best astronomical sites on Earth.

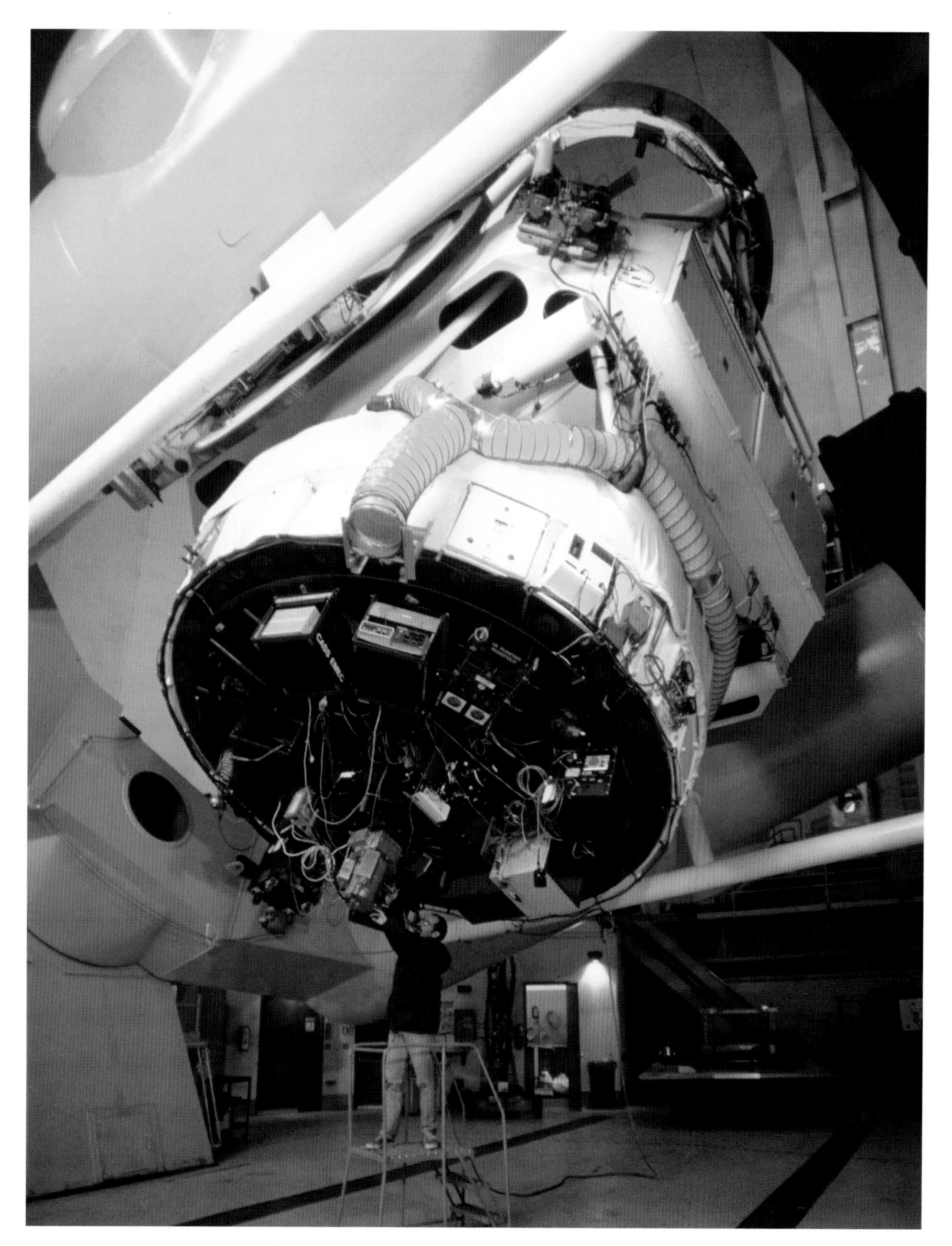

From the 1950s through the 1980s, the world's largest telescopes all looked something like the 3.6-meter Canada-France-Hawaii Telescope shown on this page. Characterized by elegant but colossal equatorial mounts and thick ceramic mirrors, these telescopes were very successful in their day (and are still used nightly around the world), but they were limited by their design to 5 meters or less in size. The new generation of telescopes (see page 159) has wider but thinner mirrors and much lighter mounts that are computer-controlled. Computers have so revolutionized astronomy, in fact, that researchers rarely look through the telescope or work in the dome during observations. Instead, they operate the instrument in comfort from a control room, above, where the object being observed is displayed on a computer monitor.

REFLECTIONS ON ASTRONOMICAL ILLUSTRATION

CHAPTER ELEVEN

Sometime during the spring of 1955, a time when I still had more than a year to wait to become a teenager, my father took me to the big city library. This was done partly to keep me quiet. I had read all the astronomy books in the library at my suburban public school and wanted more, but there were no other nearby sources.

After we passed through the library's heavy oak-and-glass doors, Dad pushed me in front of the librarian seated at the information desk. I looked at the "Quiet Please" sign, hesitated, then whispered, "Do you have any books about planets and stars?" The librarian took off her glasses, stood up, glided to the correct aisle and pointed to a shelf.

"A whole shelf of astronomy books!" I thought. This was too good to be true. And, after a closer look, I saw that indeed, it *was* too good to be true. Many were indecipherable college textbooks. The "real" astronomy books—those with titles like *A Beginner's Guide to the Stars*—totaled no more than a dozen well-worn volumes. I began to examine them as my father went over to the magazine racks to look for *Motor Trend*.

One of the books had no dustcover, just a naked navy-blue hard cover with the title *The Conquest of Space* down the spine. But when I opened it, I experienced one of those great moments of discovery—the ones that become etched in the brain for life. The full-color illustrations of alien planetary landscapes had a depth and realism unlike anything I had seen before. The illustrator was Chesley Bonestell. His paintings showed Mars as seen from its moon Phobos, Jupiter from Amalthea, Saturn from Titan and dozens more that propelled me into a new universe.

After a year of saving and pestering, I had my own copy of *The Conquest of Space*. And I was not alone in wanting a personal copy. First published in 1949, it became the best-selling astronomy book to that time and is now a prized collectors' item. *The Conquest of Space* was a unique collaboration of two visionaries: Willy Ley, a German rocket-scientist-turned-science-writer, and Chesley Bonestell, an architectural illustrator who was 56 when he painted his first piece of space art.

Ley and Bonestell produced several similar books in the 1950s, but *The Conquest of Space* was significant because it combined, for the first time, realistic and technically accurate color illustrations of planetary exploration. An entire generation of young astronomy and space-exploration buffs (like me) was profoundly influenced by this book. Thousands of scientists and engineers who years later worked on the Apollo Moon program, the Voyager flights to the giant planets and other ventures into the solar system got their first look at what it might be like in the pages of Bonestell and Ley's masterwork.

This is not mere speculation on my part. During the late 1970s and early 1980s, a period the late Carl Sagan called the "golden years of planetary exploration," I was a full-time space-science writer for newspapers and magazines, an occupation that led to hundreds of interviews with mission scientists and engineers. I learned that many of them were as profoundly influenced by Bonestell's creations as I had been. Some mentioned having seen Bonestell's work for the first time in one of the mass-circulation magazines that came into their homes in the 1950s. Most impressive of these was the series that appeared in *Collier's*

The only typical place in the cosmos is the everlasting night of intergalactic space, a place so vast and desolate that by comparison, planets and stars and galaxies seem achingly rare and lovely.

Carl Sagan

The Milky Way Galaxy rises in all its splendor above the horizon of a hypothetical planet of a star in the Large Magellanic Cloud. In this rendering, astronomical artist Adolf Schaller has placed the observer high atop a mountain poking above the clouds. The nearest edge of the Milky Way is 150,000 light-years distant. Above: View of the red-dwarf star Gliese 876 from a moon of its Jupiter-sized planet (illustration by David Egge).

Adolf Schaller's "Giant Impact on the Early Earth" displays the artist's ability to render a scene with near-photographic realism while retaining scientific accuracy. Here, a large comet slams into Earth about 100 million years after the birth of the solar system. Similar impacts on the primitive Moon created the craters that remain today, a testament to the violence of long ago.

in 1952. Written by Wernher von Braun, Willy Ley and others, the articles described the steps required for manned voyages to the Moon and Mars. Bonestell's visions of space stations and nuclear rockets to Mars were powerful, firing the imaginations of thousands of nascent rocket scientists and astronomers.

Recognizing a good idea when he saw it, Walt Disney used the *Collier's* series as the basis for several animated short features on space travel and life on other worlds, which were eventually seen by millions of people worldwide. Few 20th-century science books have influenced more people.

Bonestell was the father of modern space art, and for two decades after *The Conquest of Space* was published, he remained its most visible practitioner. A few young space artists, inspired by the master but toiling in Bonestell's shadow for years, finally emerged as masters in their own right during the 1970s and 1980s, producing stunning work portraying the wealth of new scientific discoveries during the heyday of planetary exploration. Several of them contributed to this and previous editions of *The Universe and Beyond*. But the work of Adolf Schaller, the book's major contributor, has, I think, set new standards of accuracy and excellence in astronomical illustration.

Many of the Schaller paintings in this book were completed in the early 1980s, when the artist was in his twenties (and Bonestell was in his nineties). Some of my favorite Schaller pieces are displayed in this chapter. A personal favorite of mine is the stunningly realistic work above. Schaller was a major contributor to Carl Sagan's *Cosmos* television series and was responsible for the superb renderings of Jupiter and its satellites in the film *2010*, the sequel to *2001: A Space Odyssey*. Unlike Bonestell, a professionally trained architectural illustrator, Schaller is entirely self-taught and sold his first magazine illustration at age 15.

Historical Perspective

The importance of astronomical art is, I believe, vastly underrated. Describing in words a scenario for a trip to the Moon or Mars, however detailed and well crafted, pales in comparison with the same voyages illustrated in a realistic fashion. It was the images more than the words which convinced people that space travel could happen. And after it did happen, the photographs told the story far more eloquently than did either the astronauts themselves or news anchor Walter Cronkite. For instance, the Apollo astro-

Adolf Schaller's vision of a galactic star-burst—the creation of billions of stars at a galaxy's heart one to two billion years after the Big Bang —as seen from an imaginary planet. The torrent of star births ignites at a critical point when the core of the newly forming galaxy is experiencing the maximum inflow of gas from surrounding space. Outward shock waves from ensuing supernovas collide with infalling gas to create the crescendo of cosmic fury seen here. In the foreground, a few hundred light-years from the star-burst tumult, a hypothetical planet has been subjected to millions of years of high-intensity radiation scouring that has smoothed the mammoth igneous rock formations.

nauts' photographs of earthrise over the Moon have not retreated into obscurity. Instead, they remain icons of the space age. It is no coincidence that the modern environmental movement began the same year (1970) the Apollo photographs of Earth first sank into human consciousness. This point is often overlooked by space-exploration historians.

Bonestell's 1950s depictions of Earth-orbiting space stations, based on designs by Wernher von Braun, are an example of space art's impact on our collective consciousness. These serene visions of vast metallic structures floating above the Earth's curving horizon became so implanted in the minds of a generation of engineers, designers and NASA officials that space stations became the obvious next step, even though, to this day, nobody has defined a clear mission for them. Von Braun's rationale was that a space station would be an assembly point for the construction of ships to carry crews to the Moon and Mars. This, everyone now agrees, was never a practical idea.

Another example, mentioned in a different context in Chapter 2, is the legacy of Bonestell's paintings of the surface of Mars, showing a blue sky and Arizona-like deserts. They became the "Mars of the mind" and were emulated by most other space artists before the Viking landers revealed the real rock-strewn Mars with its peach-colored sky. So ingrained was the Bonestellian Mars that the first color image of the surface of Mars released to the news media in 1976 had brown rocks and a blue sky. After checking the spacecraft color-calibration system overnight and realizing that the pinkish sky and orange surface were the true colors, the mission scientists somewhat sheepishly presented the "corrected" version the next day.

Similarly, more recent depictions of as-yet-unseen celestial vistas—pulsars, black holes, comet surfaces, alien life-forms—will remain for future generations as collective mental images of what these distant places "should" look like. To a large degree, this is a good thing. Illustrations that fire the imagination of the general population can only help to inspire future explorers.

I often think back to that day my father took me to the library and wonder what impact some of the best-illustrated modern astronomy books have on today's young people. Does it all get prefiltered by the inevitable previous exposures to the superb special effects of modern cinema, video games, and so on? Is the power of space art now swamped in a culture of information and image overload? Time will tell.

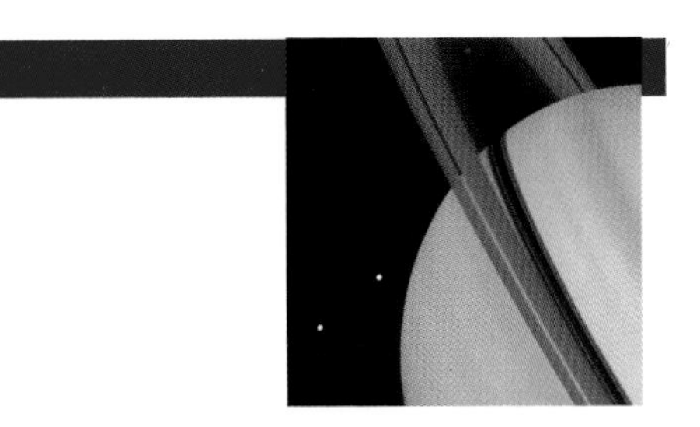

A BRIEF HISTORY OF PLANETARY EXPLORATION (1949-2004)

1949: Publication of the book *The Conquest of Space* by Chesley Bonestell and Willy Ley containing the first realistic illustrations of planetscapes and planetary exploration. This important and popular work impresses a whole generation of future astronomers and NASA designers.

1950: The feature film *Destination Moon* is released. Using imaginative special effects (for the time), it portrays the first scientifically plausible scenario for space exploration seen in a mainstream theatrical release.

1952: A series of beautifully illustrated articles about spaceflight by Wernher von Braun and others appears in *Collier's*, a mass-circulation American magazine. The articles deeply influence thousands of readers who would later pursue careers in science and engineering.

1957: Soviet Union puts into orbit the first artificial satellite, Sputnik l, on October 4. The space age is born.

1958: The first U.S. satellite, Explorer 1, is launched by Wernher von Braun's team on January 31.

1959: Soviet Union's Luna 3 circles the Moon and transmits to Earth the first photographs of the Moon's far side.

1961: Yuri Gagarin, the first person in space, makes a single orbit of Earth on April 12 in his Vostok 1 capsule. He lands safely in the Soviet Union.

1962: U.S. Mariner 2 probe passes Venus; the first successful flyby of any planet.

1965: U.S. Mariner 4 makes the first successful flyby of Mars. Sends back fuzzy pictures showing craters.

1966: Soviet Union's Luna 9 sends back the first pictures from the surface of the Moon.

1967: Soviet Venera 4 spacecraft makes the first entry into the atmosphere of another planet as it descends by parachute into the dense clouds of Venus. It does not survive to the surface.

1968: Release of the landmark movie *2001: A Space Odyssey* heightens the drama surrounding the upcoming Apollo Moon flights and romances a new generation of space-travel enthusiasts.

1968: Three astronauts in Apollo 8 are the first humans to orbit the Moon during test of Moon-flight hardware.

1969: Apollo 11 astronauts Neil Armstrong and Buzz Aldrin walk on the Moon on July 20. More than one billion people watch the event live on television.

1970: Soviet Venera 7 probe lands on Venus and transmits data from surface. This is the first successful landing on another planet.

1970: Soviet Lunokhod 1 wheeled Moon rover is the first mobile robot device on another world.

1971: Soviet Mars 3 probe lands on Mars but returns only 20 seconds of data.

1971: U.S. Mariner 9 probe orbits Mars to become the first device to orbit another planet; it maps most of the red planet.

1972: The Apollo 17 mission ends the U.S. Moon-landing program. In six landings (1969-72), 12 astronauts explored the lunar surface.

1973: U.S. Pioneer 10 is the first probe to fly by Jupiter.

1974: Mariner 10, the first spacecraft to reach Mercury, photographs the planet during flyby.

1975: Venera 9 transmits to Earth the first pictures from the surface of another planet—Venus.

1976: First pictures from the surface of Mars are transmitted to Earth by Vikings 1 and 2.

1979: Pioneer 11 is the first spacecraft to reach Saturn.

1979: Voyagers 1 and 2 fly by Jupiter and gather the first detailed photographs of the giant planet.

1980: First detailed pictures of Saturn—from Voyager 1 flyby.

1983: U.S. Pioneer 10 probe reaches a point farther from the Sun than all the planets; first human artifact to leave the solar system.

1986: Voyager 2 makes the first flyby of Uranus and sends back high-quality pictures.

1986: A flotilla of five space probes from Europe, Japan and the Soviet Union explores Halley's Comet.

1989: Voyager 2 makes the first flyby of Neptune and sends back excellent-quality pictures.

1991: On its way to Jupiter, Galileo probe takes first close-up photographs of an asteroid.

1995: Galileo becomes the first spacecraft to orbit Jupiter; over five years, thousands of images of Jupiter and its moons are gathered.

1997: Pathfinder lands on Mars and releases first roving vehicle on another planet.

2001: NEAR Shoemaker is the first spacecraft to make a "soft" landing on an asteroid.

2004: Cassini is the first spacecraft to orbit Saturn.

DISCOVERY OF THE UNIVERSE: *A Brief Chronology*

2200 B.C.: Most major constellations are developed; systematic observation of the movements of the planets and other sky motions begins in Mediterranean region, China and India.

350 B.C.: Aristotle proposes that the universe consists of concentric crystalline spheres centered on Earth; the Moon and planets are attached to the inner spheres, while the stars ride on the outer one.

260 B.C.: Aristarchus hypothesizes that Earth orbits the Sun.

200 B.C.: Eratosthenes calculates the correct diameter of Earth.

150 B.C.: Greek astronomer Hipparchus obtains positions for the 1,000 brightest stars and thereby devises the world's first accurate star chart.

150 A.D.: Ptolemy works out a complex geocentric (Earth-centered) system to explain the workings of the solar system; though incorrect, it makes reasonably accurate predictions and is accepted wisdom for the next 15 centuries.

1543: Publication of Copernicus's *Revolution of the Heavenly Spheres,* which shows that the planets orbit the Sun.

1572: Tycho Brahe observes a supernova as bright as Venus in the constellation Cassiopeia, proving that the starry vault can change.

1609: Galileo is the first to observe the Moon and other celestial objects through a telescope.

1659: Christiaan Huygens discovers the rings of Saturn.

1687: Sir Isaac Newton's theory of gravity explains the motions of celestial objects.

1750: Charles Messier begins cataloging fuzzy celestial objects that might be confused with comets—the famous *Messier Catalog*.

1755: Immanuel Kant suggests that some nebulas are galaxies.

1838: First accurate distance to a star measured.

1850: First successful astronomical photograph taken (the Moon).

1874-77: First calculations of close-to-modern values of the distances from Earth to the Sun and other planets.

1894: Percival Lowell theorizes that the "canals" of Mars are evidence of intelligent life on the nearby planet. Widely reported in the press, his assertion marks the beginning of popular interest in the possibility of extraterrestrial life that continues today.

1912: Discovery of period-luminosity relationship of Cepheid variable stars; still one of the most powerful tools for measuring distances to nearby galaxies.

1915: Einstein develops theory of general relativity.

1918: Harlow Shapley shows that the Sun is off toward the edge of the Milky Way Galaxy.

1922: Various lines of evidence strongly suggest that the so-called spiral nebulae are, in fact, remote galaxies similar to our own Milky Way Galaxy.

1925: Edwin Hubble detects Cepheid variable stars in Andromeda's "Great Nebula," proving that it is a galaxy of stars.

1929: Based on the redshift of dozens of galaxies, obtained by Edwin Hubble and Milton Humason using the Mount Wilson 100-inch telescope, astronomers deduce that on the large scale, the universe is expanding.

1940s: The theory that planets may be a natural by-product of the formation of stars gains favor.

1948: George Gamow and others propose what eventually becomes known as the Big Bang theory of the creation of the universe.

1958: First rough map of the spiral-arm structure of our galaxy developed from radio-telescope observations.

1963: Discovery of quasars.

1965: Discovery of the cosmic background radiation, the most important pillar of support for the Big Bang theory.

1967: Discovery of pulsars.

1969: Humans walk on the Moon.

1970: Discovery of Cygnus X-1, the first persuasive example of a black hole.

1980s: The development of electronic detectors many times more sensitive than photographic film revolutionizes the acquisition of astronomical images and information.

1990: Launch of the Hubble Space Telescope.

1990s: Several lines of evidence strongly suggest that the universe will continue to expand forever.

1995: First discovery of planets around other Sunlike stars.

1998: Two teams of astronomers working independently discover that the expansion of the universe is accelerating, not slowing down as expected.

1999: Realization that approximately 70 percent of the universe is "dark energy," the nature of which is unknown.

THE PLANETS AND THEIR MOONS

In this illustration, the Sun, its planets and all their known moons are shown at correct scale size. Of course, accurate scale distances cannot be depicted in an illustration like this. To do so would require a sheet of paper 20 kilometers across. The Sun and Mercury, for example, would be 60 meters apart.

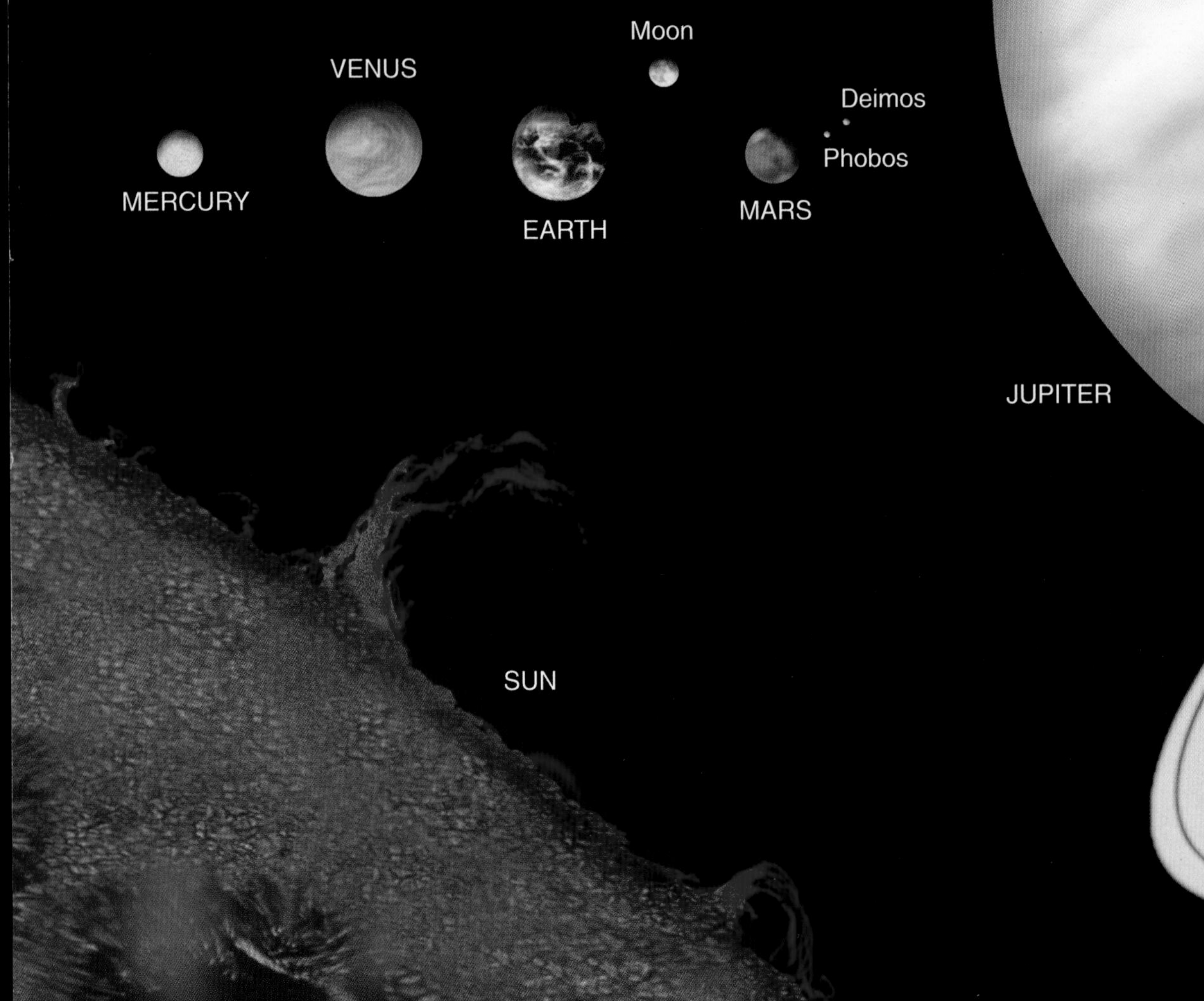

Pasiphae
Sinope
Ananke
Carme
Lysithea
Elara
Leda
Himalia
Callisto
Ganymede
Europa
Io
Thebe
Amalthea
Adrastea
Metis
Phoebe
Iapetus
Hyperion
Titan
Rhea
Helene
Dione
Calypso
Tethys
Enceladus
Telesto
Janus
Mimas
Pandora
Epimetheus
Atlas
Prometheus
Pan
Sycorax
Caliban
Oberon
Titania
Umbriel
Ariel
Miranda
Puck
Belinda
Rosalind
Portia
Juliet
Desdemona
Cressida
Bianca
Ophelia
Cordelia
URANUS
SATURN
MOON UPDATE
In recent years, searches with the world's largest telescopes have revealed additional small moons orbiting at large distances from the four giant planets. See "Known Moons" column in the table on the next page for totals as of 2004.
NEPTUNE
Naiad
Thalassa
Despina
Galatea
Larissa
Proteus
Triton
Nereid
Charon
PLUTO

DATA

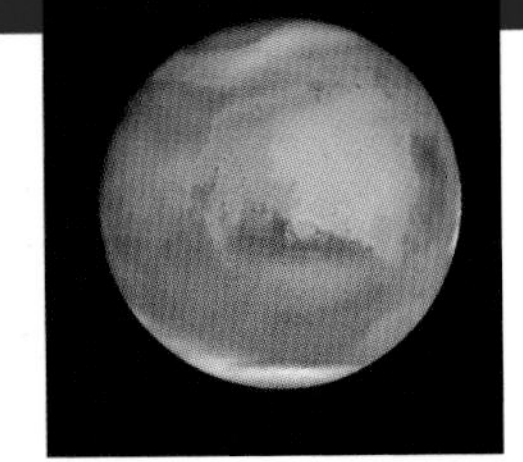

THE SUN AND ITS PLANETS

Object	Diameter (Earth=1)	Mass (Earth=1)	Density (water=1)	Distance from Sun (Earth's distance=1)	Rotation period*
SUN	109.1	332,946	1.41	—	25 to 35 days
MERCURY	0.382	0.055	5.43	0.387	176 days
VENUS	0.949	0.815	5.24	0.723	117 days
EARTH	1.000	1.000	5.52	1.000	24 hours
MARS	0.532	0.107	3.94	1.524	24h 39m
JUPITER	11.19	317.8	1.33	5.20	9h 50m
SATURN	9.41	95.2	0.77	9.54	10h 39m
URANUS	4.01	14.5	1.27	19.18	17h 14m
NEPTUNE	3.88	17.2	1.64	30.06	16h 06m
PLUTO	0.19	0.0020	1.9	29.6 to 49.3	6d 9.3h

Object	Revolution period (length of year)	Average amount of sunlight received (Earth=1)	Surface gravity† (Earth=1)	Known moons	Axis inclination	Minimum light-time from Earth
SUN	—	—	27.9	—	7.3°	8.3 min.
MERCURY	88.0 days	6.6	0.38	0	0.0°	4.5 min.
VENUS	224.7 days	2.2	0.91	0	177.3°	2.3 min.
EARTH	365.3 days	1.0	1.00	1	23.4°	—
MARS	687.0 days	0.44	0.38	2	25.2°	3.1 min.
JUPITER	11.86 years	0.037	2.69	61	3.1°	33 min.
SATURN	29.46 years	0.011	1.19	31	26.7°	66 min.
URANUS	84.0 years	0.0028	0.91	22	97.9°	2.5 hours
NEPTUNE	164.8 years	0.0011	1.19	11	28.8°	4.0 hours
PLUTO	247.7 years	0.001 to 0.0004	0.06	1	120°	3.9 hours

*Rotation period of the Sun varies with latitude; for the planets, the figure given is period from one sunrise to the next at the equator.
†For Sun and giant planets, "surface" refers to *visible* surface (cloud tops and solar photosphere).
Earth's diameter is 12,756 kilometers; its distance from the Sun is 149.6 million kilometers.

MAJOR PLANETARY SATELLITES

Planet	Satellite	Diameter (Moon=1*)	Distance from planet surface in planetary radii	Orbit period (days)	Mass (Moon=1)	Surface gravity (as a percent of Earth's)
EARTH	Moon	1.00	58.75	27.32	1.00	16
JUPITER	Io	1.04	4.89	1.77	1.21	17
	Europa	0.90	8.33	3.55	0.65	14
	Ganymede	1.51	15.90	7.16	2.01	19
	Callisto	1.38	25.20	16.69	1.47	17
SATURN	Mimas	0.11	2.10	0.94	0.0005	2
	Enceladus	0.14	2.94	1.37	0.001	2
	Tethys	0.30	3.89	1.89	0.011	4
	Dione	0.32	5.27	2.74	0.014	4
	Rhea	0.44	7.72	4.52	0.034	6
	Titan	1.48	19.24	15.95	1.83	19
	Hyperion	0.07	23.55	21.28	?	1
	Iapetus	0.42	58.0	79.33	0.026	5
	Phoebe	0.06	213.8	550.46	?	1
URANUS	Miranda	0.14	4.10	1.41	0.001	2
	Ariel	0.34	6.52	2.50	0.01	6
	Umbriel	0.34	7.58	4.14	0.01	5
	Titania	0.46	16.17	8.71	0.12	8
	Oberon	0.45	21.98	13.46	0.11	8
NEPTUNE	Triton	0.78	22.08	5.88	0.31	13
PLUTO	Charon	0.40	17	6.39	0.025	3

*Moon's diameter is 3,476 kilometers.

THE NEAREST STARS

Name	Distance from Sun in light-years	Type	Mass (Sun=1)	Diameter (Sun=1)	Luminosity (Sun=1)	Remarks
SUN	—	main sequence G2	1	1	1	nearest star
PROXIMA CENTAURI*	4.22	red dwarf M5	0.11	0.11	0.00006	nearest star to solar system
ALPHA CENTAURI A	4.40	main sequence G2	1.1	1.2	1.3	A to B=24 AU
ALPHA CENTAURI B	4.40	main sequence K1	0.9	1.0	0.36	AB to Proxima=20,000 AU
BARNARD'S STAR	5.94	red dwarf M4	0.15	0.12	0.00044	
WOLF 359	7.7	red dwarf M6	0.10	0.1	0.00002	
BD+36°2147A	8.2	red dwarf M2	0.35	0.3	0.0052	A to B=0.07 AU
L726-8 A	8.4	red dwarf M6	0.44	0.1	0.00006	A to B=11 AU
L726-8 B	8.4	red dwarf M6	0.35	0.1	0.00004	
SIRIUS A	8.6	main sequence A1	2.3	1.8	23	brightest within 25 light-years
SIRIUS B	8.6	white dwarf	1.0	0.02	0.003	A to B=20 AU
ROSS 154	9.7	red dwarf M4	0.15	0.12	0.0005	
ROSS 248	10.4	red dwarf M5	0.13	0.1	0.0001	
EPSILON ERIDANI A	10.5	main sequence K2	0.76	0.9	0.3	
ROSS 128	10.9	red dwarf M4	0.15	0.12	0.00035	
61 CYGNI A	11.4	main sequence K4	0.6	0.7	0.08	A to B=85 AU
61 CYGNI B	11.4	main sequence K5	0.5	0.6	0.04	
EPSILON INDI	11.8	main sequence K3	0.6	0.7	0.13	
BD+43°44 A	11.2	red dwarf M1	0.3	0.3	0.006	A to B=156 AU
BD+43°44 B	11.2	red dwarf M4	0.15	0.13	0.0004	
L789-6	11.2	red dwarf M6.5	0.13	0.1	0.0002	
PROCYON A	11.4	main sequence F5	1.8	1.7	7.6	A to B=16 AU
PROCYON B	11.4	white dwarf	0.6	0.01	0.0005	
TAU CETI	11.9	main sequence G8	0.8	0.9	0.44	first object searched for ET radio signals

Distances are known within 0.1 light-year; most masses and diameters are estimates; all listed objects are invisible to unaided eye except: Sun, Alpha Centauri (A & B appear as one star), Sirius A, Epsilon Eridani A, 61 Cygni A, Epsilon Indi, Procyon A and Tau Ceti.

One light-year equals 9.46 trillion kilometers.

AU (astronomical unit)=the distance from Earth to Sun; 149.6 million kilometers.

*Proxima is the outer component of the Alpha Centauri triple-star system.

A SELECTION OF PLANETS OF OTHER STARS *(in order of increasing mass)*

As of 2004, more than 120 extrasolar planets are known

Star (with distance)	Mass of planet* (Jupiter masses)	Distance from star (AU)	Orbital period	Remarks
HD76700 (197 ly)	0.20	0.049	3.97 days	least massive extrasolar planet known (as of 2004)
51 Pegasi (50 ly)	0.47	0.052	4.32 days	first extrasolar planet discovered (1995)
Upsilon Andromedae (45 ly) (3-planet system)	0.69 1.19 3.75	0.059 0.829 2.53	4.62 days 241 days 3.52 years	three planets
55 Cancri (55 ly) (3-planet system?)	0.84 0.21? 4.05	0.11 0.24? 5.9	14.65 days 44 days? 14.8 years	outer planet in Jupiterlike orbit; middle planet uncertain
Epsilon Eridani (10.5 ly)	0.86	3.3	6.85 years	nearest Sunlike star with known planet
OGLE-TR-56 (5,000 ly)	1.45	0.0225	1.2 days	shortest planetary orbital period known; discovered using planetary-transit technique
16 Cygni B (72 ly)	1.69	0.6 to 2.7	2.19 years	very eccentric orbit
47 Ursae Majoris (44 ly) (2-planet system)	2.41 0.76	2.10 3.73	3.00 years 7.10 years	star is an estimated 2.5 billion years older than the Sun
14 Herculis (60 ly)	4.74	2.80	4.92 years	very eccentric orbit
70 Virginis (59 ly)	7.44	0.48	116.7 days	very eccentric orbit

*Objects more massive than 13 Jovian masses are considered brown dwarfs.

SOME WELL-KNOWN BRIGHT STARS

Name	Apparent brightness*	Luminosity (Sun=1)	Distance in light-years	Diameter (Sun=1)	Mass (Sun=1)	Type and spectrum	Remarks
SIRIUS A	100	23	8.6	1.8	2.3	main sequence A1	white-dwarf companion
CANOPUS	52	11,000	310	30	6	giant A9	
ALPHA CENTAURI A	35	1.3	4.4	1.2	1.1	main sequence G2	triple star system
ARCTURUS	29	115	37	23	4	giant K1	nearest giant
VEGA	26	55	25	3	3	main sequence A0	
CAPELLA A & B	25	90 & 70	42	13 & 7	3 & 2.5	giant G2 & G6	tight binary
RIGEL	24	55,000	770	50	20	supergiant B8	tight binary plus distant pair
PROCYON A	20	7	11.4	2.2	1.8	main sequence F5	white-dwarf companion
BETELGEUSE	14	8,000 to 14,000	430	500 to 800	18	supergiant M2	slowly varies in brightness and size
ALDEBARAN	12	125	65	45	4	giant K5	less massive companion
ANTARES	11	9,000	600	300	10	supergiant M1	main-sequence B3 companion

*As seen without optical aid in the Earth's night sky; figure given is a percent, in comparison with Sirius, which is arbitrarily given a value of 100.
Distances to stars within 200 light-years are known to within a few percent. Diameters and masses of most stars are estimates.

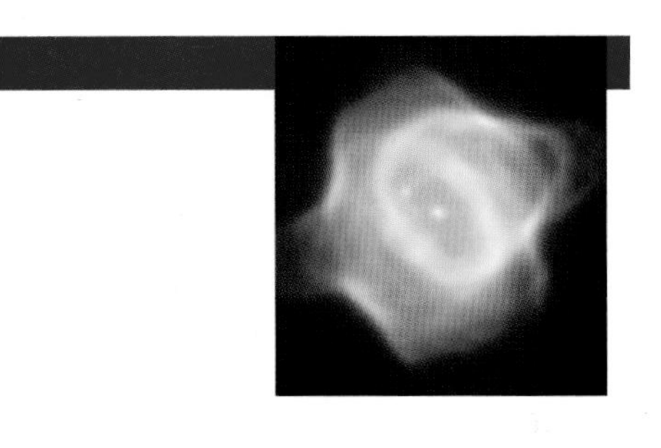

SOME CHARACTERISTICS OF MAIN-SEQUENCE STARS

Mass	Spectral type	Luminosity (Sun=1)	Diameter (Sun=1)	Central density (water=1)	Lifetime on main sequence (millions of years)
30	O8	100,000	15	3	4
15	B1	20,000	10	6	11
10	B5	5,000	5	9	20
5	B8	500	3	20	75
2	A6	17	1.7	70	800
1.5	F3	5	1.3	85	1,800
1	G2	1	1	90	10,000
0.5	K8	0.03	0.7	80	100,000
0.1	M7	0.0001	0.1	60	1,000,000

Mass	Death process	Mass of final remnant star	Probable form of remnant
30	supernova	5 to 20	black hole
15	supernova	1.4 to 10	black hole or neutron star
10	supernova	0.5 to 4	neutron star or white dwarf
5	gradual mass loss and collapse	less than 1.4	white dwarf
2	gradual mass loss and collapse	less than 1.4	white dwarf
1.5	gradual mass loss and collapse	less than 1.4	white dwarf
1	minor mass loss and collapse	about 0.9	white dwarf
0.5	gradual cooling	same	white dwarf
0.1	gradual cooling	same	brown dwarf

LIFE HISTORY OF A MASSIVE STAR *(15-solar-mass example)*

0 to 10,000 years	contraction of nebula, collapse of cloud and formation of star, leading to ignition of thermonuclear reactions at core
10,000 to 10,000,000 years	main-sequence star: combustion of hydrogen in the core; luminosity about 30,000 times the Sun's
next 200,000 years	beginning of helium combustion in core; star begins to expand from 10 to 30 solar diameters
next 700,000 years	continued expansion to 50 solar diameters; star begins to pulsate as a Cepheid variable
next 700,000 years	irregular variability sets in; expansion continues to 100 solar diameters
next 30,000 years	expands to red supergiant like Betelgeuse, 500 times the Sun's diameter; luminosity fairly constant throughout all above phases
sudden onset	supernova; total life span about 11 million years

LIFE HISTORY OF THE SUN

Event	Age (millions of years)	Luminosity (present Sun=1)	Diameter (present Sun=1)
Contraction from nebula into protostar	0	100	50
Hot core forms from contraction	1	20	20
		Main-Sequence Life Begins	
Protostar contraction ends; nuclear fusion of hydrogen begins	70	0.6	1.0
Today	4,600	1.0	1.0
Hydrogen begins to be depleted at Sun's core	7,000	1.4	1.2
Hydrogen fusion requires higher temperatures due to helium buildup	9,000	2	1.5
Hydrogen "burning" moves to a shell around helium core	10,000	4	3
		Main-Sequence Life Ends	
First red-giant stage reaches maximum, and helium core ignites	10,600	1,500	50
Helium fusion in core nears maximum	10,630	100	10
Final red-giant stage begins	10,650	1,000	100
Final red-giant stage reaches maximum as helium fusion shifts to shell around core	11,000	10,000	400
Sun sheds matter as variable or planetary nebula	11,000	variable	contracting
White dwarf forms in 75,000 years	11,000	1/300	1/100
White dwarf cools to black dwarf	50,000?	0	1/100

FURTHER READING

Although hundreds of popular-level astronomy books are published in the English language each year, just a handful rise to the top to become worthy of a place on every astronomy enthusiast's bookshelf. Here's a selection:

Solar System

Beyond: Visions of the Interplanetary Probes by Michael Benson (Abrams, 2003) is the most spectacular collection of spacecraft images of the planets and their moons ever compiled.

Carl Sagan's Universe (Cambridge, 1997) is an eclectic collection of entertaining essays on planetary exploration and related subjects that were presented at a symposium in Sagan's honor in 1994.

Distant Wanderers: The Search for Planets Beyond the Solar System by Bruce Dorminey (Copernicus Books, 2002) is probably the best of a dozen or so books that have been written about the detection of planets orbiting other stars.

The New Solar System, 4th edition, edited by J. Kelly Beatty et al. (Sky Publishing, 1999) is, by a wide margin, the best single reference book on the solar system currently available. Comprehensive and superbly illustrated.

Pale Blue Dot by Carl Sagan (Random House, 1994) is a masterful work by the greatest astronomy communicator of the 20th century.

A Traveler's Guide to Mars by William K. Hartmann (Workman Publishing, 2003) is a lavishly illustrated, full-color guide to the surface features of our neighbor world. Highly recommended.

Cosmology

Two books I enthusiastically recommend in this category are *The Whole Shebang* by Timothy Ferris (Simon & Schuster, 1997) and *Coming of Age in the Milky Way*, also by Timothy Ferris (Morrow, 1988). Ferris is an outstanding astronomy writer who deftly handles the full range of modern and historical cosmology.

Before the Beginning by Martin Rees (Addison Wesley, 1997) is an authoritative "state of the universe" report written by Britain's Astronomer Royal.

Black Holes and Time Warps: Einstein's Outrageous Legacy by Kip Thorne (Norton, 1994) is one of the few 600-page books in my astronomy library that I wish were *longer*. Thorne, a Caltech physicist and one of the world's leading authorities on black holes, does a splendid job of explaining the bizarre nature of the gravitational abyss of black holes and related subjects, such as wormholes, neutron stars, relativity and the gravitational warping of space and time.

The Extravagant Universe by Robert P. Kirshner (Princeton University Press, 2002) is mainly devoted to the story of the discovery of the accelerating universe in the late 1990s. Kirshner, a supernova expert, was one of the astronomers involved in the process.

The Fabric of the Cosmos by Brian Greene (Knopf, 2004), a physicist and best-selling author, takes the reader on a journey into the realm of modern concepts of time and string theory to explain the Big Bang itself and the behavior of everything from the smallest particle to the largest black hole. Also see Greene's earlier book *The Elegant Universe* (Vintage Press, 2000).

Lonely Hearts of the Cosmos by Dennis Overbye (HarperCollins, 1991) is one of the finest astronomy books ever written. It is a compelling narrative that describes how leading cosmologists conduct their work.

Extraterrestrial Life

Cosmic Company: The Search for Life in the Universe by Seth Shostak and Alex Barnett (Cambridge, 2003) details the background to the radio-search strategy for the quest for extraterrestrial life. Entertainingly written at a high school level, with color illustrations on virtually every page. Shostak is a senior astronomer at the SETI Institute.

Extraterrestrials: A Field Guide for Earthlings by Terence Dickinson and Adolf Schaller (Firefly, 1995) is a heavily illustrated introduction to the scientific quest for life beyond Earth for younger readers.

Life on Other Worlds by Steven J. Dick (Cambridge, 1998) is a definitive work that covers every aspect of the extraterrestrial-life debate, from the "canals" of Mars to UFOs to radio-telescope searches. The ultimate reference on the subject.

Planetary Dreams by Robert Shapiro (Wiley, 1999) is an authoritative and well-written look at the quest for life beyond Earth.

Rare Earth by Peter D. Ward and Donald Brownlee (Copernicus Books, 2000) is a controversial and fascinating look at many of the standard arguments about the number of habitable worlds in the cosmos. Also by the same authors, *The Life and Death of Planet Earth* (Times Books, 2002), subtitled "How the New Science of Astrobiology Charts the Ultimate Fate of Our World."

Where Is Everybody? by Stephen Webb (Copernicus Books, 2002) offers 50 solutions to the Fermi paradox, which states the essential question about extraterrestrial life—where is everybody? Webb's book is a refreshing new way of looking at some of the well-worn arguments about the possibility of extraterrestrial life. Entertaining and enlightening.

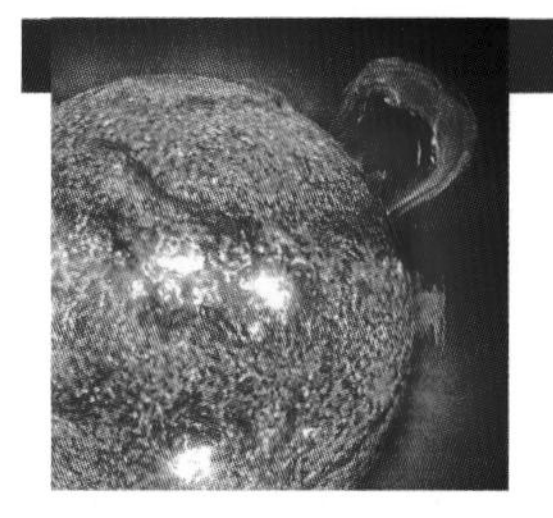

ILLUSTRATION CREDITS

FRONT COVER: Jean-Charles Cuillandre
BACK COVER: Hubble Space Telescope

p.3 Space Telescope Science Institute; p.6 European Southern Observatory; p.7 NASA; p.8 NASA/JPL; p.9 Tony Hallas

CHAPTER 1: p.10 Alan Dyer; p.11 NASA; p.12-13 Illustration by John Bianchi; p.14 NASA; p.15 (left) NASA; p.15 (right) Jack Newton; p.16 NASA; p.17 (left) Canada-France-Hawaii Telescope; p.17 (right) Terence Dickinson; p.18 Alan Dyer; p.19 Space Telescope Science Institute

CHAPTER 2: p.20 Illustration by Adolf Schaller; p.21 NASA; p.22 (left) Alan Dyer; p.22 (right) Illustration by Adolf Schaller; p.23 (left) Illustration by Adolf Schaller; p.23 (right) Space Telescope Science Institute; p.24 (both) NASA/JPL; p.25 (all) NASA/JPL; p.26 NASA/JPL; p.27 NASA/MSSS; p.28 NASA/JPL; p.29 (left) ESA; p.29 (right) NASA/JPL; p.30 (both) NASA/JPL; p.31 NASA/JPL; p.32 (all) NASA/JPL; p.33 (both) NASA/JPL; p.34 (all) NASA/JPL; p.35 NASA/JPL; p.36 NASA; p.37 Terence Dickinson; p.38 NASA/JPL; p.39 (both) NASA/JPL

CHAPTER 3: p.40 NASA/JPL; p.41 NASA/JPL; p.42 (top) Illustration by Adolf Schaller; p.42 (bottom) NASA/JPL; p.43 Illustration by Adolf Schaller; p.44 Illustration by Adolf Schaller; p.45 (top) Illustration by Adolf Schaller; p.45 (center and bottom) NASA/JPL; p.46 (both) NASA/JPL; p.47 Space Telescope Science Institute; p.48 NASA/JPL; p.49 NASA/JPL; p.50 (both) Darryl Archer; p.51 Illustration by Don Davis; p.52 (left) Cornell University/Caltech; p.52 (right) Hubble Space Telescope; p.53 (all) NASA/JPL

CHAPTER 4: p.54 NASA/JPL; p.55 NASA/JPL; p.56 (both) NASA/JPL; p.57 (all) NASA/JPL; p.58 (bottom left) Illustration by David Egge; p.58 (top left, top right and bottom right) NASA/JPL; p.59 (top) Illustration courtesy NASA/JPL; p.59 (bottom) NASA/JPL; p.60 (all) NASA/JPL; p.61 (both) NASA/JPL; p.62 (left, top and bottom) European Southern Observatory; p.62 (right) Illustration courtesy NASA; p.63 Illustration by Adolf Schaller; p.64 (all) NASA/JPL; p.65 (left) Illustration courtesy NASA; p.65 (right) Space Telescope Science Institute; p.66 (top left) European Space Agency; p.66 (bottom left) Jerry Lodriguss; p.66 (right) Illustration courtesy Caltech; p.67 Terence Dickinson; p.68 (left) NASA/JPL; p.68 (right) Terence Dickinson; p.69 (both) NASA/JPL; p.70 (left) Illustration by Robbie Cooke; p.70 (top right) NASA illustration by Adolf Schaller; p.70 (bottom right) Illustration by Robbie Cooke; p.71 Illustration courtesy NASA

CHAPTER 5: p.72 Anglo-Australian Telescope Board/Royal Observatory, Edinburgh; p.73 Space Telescope Science Institute; p.74 Illustration by Adolf Schaller; p.75 (left) Illustration by Michael Carroll; p.75 (right) Illustration courtesy NASA; p.76 Illustration by David Egge; p.77 (left) Illustration by Victor Costanzo; p.77 (right) Illustration by David Egge; p.78 (left) Illustration courtesy NASA; p.78 (right) Space Telescope Science Institute; p.79 Royal Swedish Academy of Sciences; p.80 European Southern Observatory; p.81 (all) Space Telescope Science Institute; p.82 NOAO/AURA/NSF; p.83 Terence Dickinson; p.84 Robert Gendler; p.85 Canada-France-Hawaii Telescope; p.86 Space Telescope Science Institute; p.88 Tony Hallas; p.89 (top) Canada-France-Hawaii Telescope; p.89 (bottom) Illustration by David Egge; p.90 Illustration by Adolf Schaller; p.91 (left) Space Telescope Science Institute; p.91 (right) Anglo-Australian Telescope Board; p.92 Tony Hallas; p.93 (all) Space Telescope Science Institute; p.94 Illustration by Adolf Schaller; p.95 Illustration by Adolf Schaller; p.96 Illustration by Adolf Schaller

CHAPTER 6: p.98 Anglo-Australian Telescope Board; p.99 Japanese National Observatory; p.100 (all) Illustrations by Brian Sullivan; p.101 (both) Space Telescope Science Institute; p.102 Anglo-Australian Telescope Board/Royal Observatory, Edinburgh; p.103 (top) European Southern Observatory; p.103 (bottom) Mike Mayerchak; p.104 Space Telescope Science Institute; p.105 Space Telescope Science Institute; p.107 Space Telescope Science Institute; p.108 Illustration by Adolf Schaller; p.109 (left) Terence Dickinson; p.109 (right) NOAO/AURA/NSF/WIYN Consortium; p.110 NASA; p.111 Illustration by Adolf Schaller; p.112 European Southern Observatory; p.113 (left) European Southern Observatory; p.113 (right) Hubble Space Telescope; p.114 European Southern Observatory; p.115 (left) Illustration by David Egge; p.115 (right) Illustration courtesy NASA; p.116 Hubble Space Telescope; p.117 Hubble Space Telescope

CHAPTER 7: p.118 Hubble Space Telescope; p.119 NOAO/AURA/NSF; p.120 M. Seldner, B. Siebers, Edward Groth, P.J.E. Peebles; p.121 Lowell Observatory; p.122 Illustration by Victor Costanzo; p.123 The 2dF Galaxy Redshift Survey ©AAO; p.124 (both) Hubble Space Telescope; p.125 Illustration courtesy NASA; p.126-127 Illustration by Adolf Schaller; p.128 NASA/WMAP; p.129 Illustration courtesy NASA; p.131 Hubble Space Telescope; p.132-133 Hubble Space Telescope; p.134 Illustration by Adolf Schaller; p.135 Illustration by Adolf Schaller

CHAPTER 8: p.136 Illustration by Adolf Schaller; p.137 National Radio Astronomy Observatory; p.138 (both) Illustrations by Adolf Schaller; p.139 (left) Illustration by Don Davis; p.139 (right) Illustration by David Egge; p.140-141 George Greaney; p.142 Alan Dyer; p.143 (left) Illustration by David Egge; p.143 (right) Illustration by Adolf Schaller; p.144 NASA/JPL; p.145 (both) Illustrations by David Egge; p.146 Terence Dickinson; p.147 Illustration by Don Davis

CHAPTER 9: p.148 Anglo-Australian Telescope Board; p.149 Tony Hallas; p.150 Terence Dickinson; p.151 University of Arizona; p.152 Terence Dickinson; p.153 European Southern Observatory

CHAPTER 10: p.154 Hubble Space Telescope; p.155 Hubble Space Telescope; p.156 (all) Hubble Space Telescope; p.157 (left) Hubble Space Telescope; p.157 (right) Hale Observatories; p.158 (both) Gemini Telescope; p.159 (top left and right) Gemini Telescope; p.159 (bottom left) European Southern Observatory; p.160 (left) Illustration courtesy NASA; p.160 (right) Illustration courtesy European Southern Observatory; p.161 (both) Terence Dickinson

CHAPTER 11: p.162 Illustration by Adolf Schaller; p.163 Illustration by David Egge; p.164 Illustration by Adolf Schaller; p.165 Illustration by Adolf Schaller

ASTRONOMICAL DATA: p.166 NASA/JPL; p.167 NASA; p.168-169 Illustration by John Bianchi; p.170 Hubble Space Telescope; p.171 Hubble Space Telescope; p.172 Alan Dyer; p.174 Space Telescope Science Institute; p.175 Space Telescope Science Institute; p.176 Terence Dickinson; p.177 NASA; p.178 Terence Dickinson; p.180 (left) Terence Dickinson; p.180 (right) Bernard Clark

INDEX